Molecules to Form: A Structural View

Nain

Contents

A Supplements

B MATLAB Coding

Abbreviations

Nomenclature

Curriculum vitae

Chapter 1

Introduction

1.1 The aim of *structural biology*

The objective of *structural biology* is within the meaning of the words themselves. "Structural" originates from the Latin word *structura* meaning the assembly, order or building of an object [1]. "*Biology*" is the science of living nature like plants and animals, and the laws of the course of life [2]. Thus, *structural biology* is the investigation of the assembly of living beings. Within a cell of any living organism, biochemical reactions and interactions, e.g. metabolism or growth, to sustain life take place. Millions of macromolecular machines operate these biochemical processes within the cell [3]. Often mentioned as the motors of the cells, the macromolecular machines occur as either proteins or RNA. Biochemical processes such as the duplication of genetic material, protein synthesis or protein degradation are carried out by these proteins or RNA. Hereby, both structures are able to assemble to multi-component protein complexes (see Figure 1.1). A human organism, e.g., contains approximately 10,000 to 20,000 differently shaped proteins and protein complexes [4]. Similarly to the design of beverage crates, which are used to transport multiple bottles, the assembly of a protein serves its particular purpose in the human organism [5]. If a protein is incorrectly assembled, the functionality of the protein is most likely disturbed. The human organism responds to this defect potentially by malfunctioning. Diseases such as Parkinson's disease or Alzheimer's disease are related to misfolded protein complexes [4]. Here, with the understanding of the assembly of the misfolded and the regular structure the knowledge about the disease is expanded, which further aims to find strategies to prevent the misfolding process.

In general, the research field of *Structural biology* focuses on the comprehension of the assembly and the related operating principle of proteins and protein complexes in order to prevent or treat dysfunctions in the living body [5].

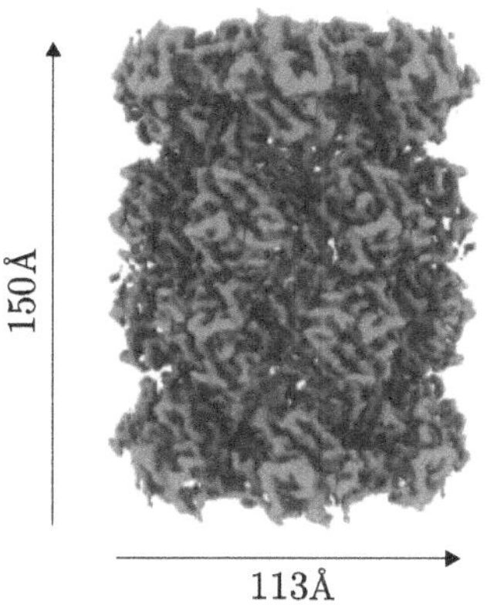

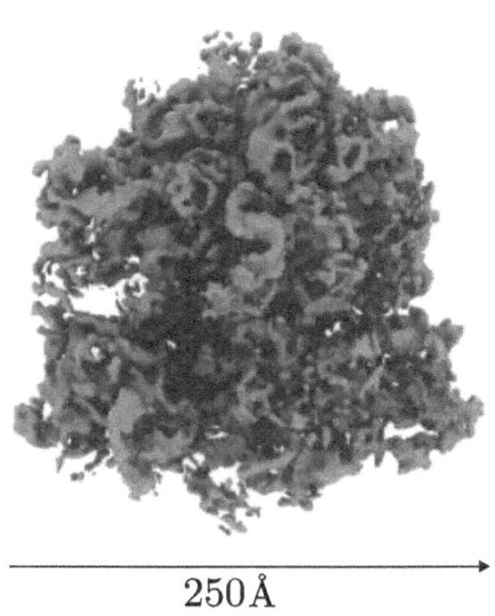

(a) Thermoplasma acidophilum 20S protea-some 3.3 Å [6]

(b) Structure of SelB-Sec-tRNASec bound to the 70S ribosome in the initial binding state (IB) with a reported resolution 5.3 Å [7]

Figure 1.1: Exemplary protein complexes Here, two cryo-EM resolved structures are shown. The T20S proteasome is a symmetric protein complex, which degrades other proteins and protein complexes. The ribosome, an asymmetric protein-RNA complex, reads the genetic code to build other protein molecules. Both structures were processed with methods of single particle analysis cryo-EM.

The assembly of a protein complex Proteins and protein complexes are unique assemblies. A protein complex is a structural formation of multiple different types of proteins or multiple copies of the identical protein. The ribosome, the protein-RNA complex in Figure 1.1b, which synthesizes other proteins and protein complexes, is the assembly of 1/3 of proteins and 2/3 of RNA. Hereby, a unique arrangement of different amino acids defines the proteins. In all, there exist unique proteinogenic amino acids [3], which fold, bend and twist to build a stable assembly. Carbon, hydrogen, oxygen, and nitrogen are the main atomic components of an amino acid sequence. The smallest atom is the hydrogen atom which has a diameter of around 0.74 Å ($Å = 10^{-10}$ m). The ribosome which is assembled of a variety of these atoms has a diameter of about 250 Å.

Proteins and protein complexes are dynamical objects in the cell. One particular structural folding of a protein complex is called conformation [3]. Through chemical activation the amino acids rearrange such that the protein complex can move from one conformation to another conformation of the complex [3]. These dynamic changes in conformation are essential for the functionality of the protein complex. As a consequence, one protein complex is capable to appear in multiple different conformations, which all serve the operating principle in the cell.

Proteins and protein complexes occur in different geometrical representations. Specific geometric arrangements can be divided into subgroups of identical shape. A protein complex, which contains two or more of these identical structural components, is called symmetric. Depending on the position and the number of the identical shapes the symmetry type and order are determined. With increasing number of these symmetric units the order of symmetry increases. The Thermoplasma acidophilum 20S proteasome in Figure 1.1a, which degrades proteins and protein complexes in thermophilic bacteria, is symmetric of higher order. In comparison, the ribosome in Figure 1.1b is asymmetric.

The structures of the protein complexes, the T20S proteasome and the ribosome, are chosen to be the models used in this **book** because both complexes have been published by several structural methods in the past. Accordingly, these structures are known up to high resolutions. The maps are accessible in the Protein Data Bank (RCSB PDB). The prior knowledge about these maps gives the opportunity to cross-validate the results of the latter presented experiments with published data.

1.1.1 Methods of structural biology

A protein complex is absolute transparent to light stated Zernike [8] in 1942. By Abbe's diffraction limit light waves with the smallest wavelength, 380 nm, are able to resolve two points in the object distanced by about 200 nm. Protein complexes are small objects (see T20S proteasome 15 nm and 11.3 nm in Figure 1.1a). The light microscope does not have the required resolution power to visualize the atomic features of the protein complex. Thus, *Structural biology* needs different techniques based on image sources such as electrons. In comparison to light, the wavelengths of electron waves (see Figure 1.2) are smaller and depend on the acceleration voltage as defined by DeBroglie. Theoretically, electrons are capable to resolve features up to atomic resolution level of the imaged material. Methods such as electron microscopy (EM) but also Nuclear Magnetic Resonance Spectroscopy (NMR spectroscopy) and X-Ray Diffraction Crystallography (XRC) were developed to visualize protein and protein complexes in the research field of *Structural biology*.

Electrons	X-rays	Visible light	Radio waves
100keV - 3.70pm 300keV - 1.96pm	0.01nm - 10nm	380nm - 750nm	1mm - 100km

Figure 1.2: Wavelengths of imaging sources Electron waves dependent on the accelerating voltage. The higher this voltage is the smaller is the wavelength. The wavelength of an electron is much shorter than the wavelength of photons in visible light. Radio waves, e.g., can travel up to 100 km until reaching a full circle (see 2.1). Conversion 1 nm is 1000 pm.

X-Ray Diffraction Crystallography   XRC is one of the oldest and robustest methods in *structural biology* [9]. Roentgen was awarded the Nobel Prize for discovering X-radiation back in 1901. 85 percent of the published protein structures as seen in Figure 1.3 are solved by XRC [5]. The aim of XRC is to reconstruct the crystallized protein structure by imaging a crystal. Many proteins of identical conformation and composition are crystallized to form an aligned lattice. The diffraction pattern of the crystal is detected. The wavelengths of X-rays vary between a few nanometers. Imaging with X-rays recovers the amplitudes of the diffracted wave functions but the phase information of the scattered ray is lost. One advantage is that XRC is not limited by the size of the protein complex [10], though crystallizing a protein complex has two disadvantages. Naturally, the protein complexes are in an aqueous solution [11]. Thus, through the crystallization process the complexes lose their conformational variability [11]. This is a disadvantage of the XRC. Another disadvantage is that molecules, especially membrane proteins, do not always crystallize [9].

Nuclear Magnetic Resonance Spectroscopy   The basic principle of NMR spectroscopy was discovered by I. I. Rabi of Columbia University. He successfully measured nuclear magnetic interactions in 1938 [12]. It took until 1946 to perform the first NMR spectroscopy by F. Block and E. M. Purcell. The aim of NMR spectroscopy is to define the distance between bonded atoms within a protein and its different conformation. This method benefits from the magnetic properties of an atom. On applying a strong external magnetic field the atoms are excited [13]. The energy, which will be absorbed, and the intensity of the signal are in relation to the strength of the magnetic field and hence, give knowledge about structural details of the protein. One asset of the NMR spectroscopy is the possibility to study proteins in a liquid solution so that the proteins occur in their near-native state. Additionally, it gives the possibility to investigate time-resolved states. The disadvantage of the NMR spectroscopy is that it is size restricted and therefore, mainly studies smaller proteins [11].

Electron microscopy   EM started back in 1931, when Ernst Ruska and his colleague Max Knoll were able to build the first TEM [14]. EM is one of the youngest methods in *structural biology*. The first published structures in the RCSB PDB are from 1997. The aim of EM is to detect the interaction of electrons with biological matter. In general, electrons are negatively charged subatomic particles and have a small wavelength to overcome the diffraction limitation. EM is split into different subgroups. There are different microscope techniques called Transmission Electron Microscope, Scanning Electron Microscope (SEM) and a more recent technique called Scanning Transmission Electron Microscope (STEM). Furthermore, to prepare and process biological data there exist single particle analysis, electron cryo-tomography [15] and electron crystallography [16]. Electron cryo-tomography focuses on the study of larger objects such as cells. In comparison, SPA cryo-EM is the imaging of non-crystallized protein complexes in cryogenic environment and the following processing of thousands of these identical particle projection images [17]. One main advantage of EM is the image acquisition of protein complexes in their non-crystallized state. Another advantage is the preserved phases of the projection images.

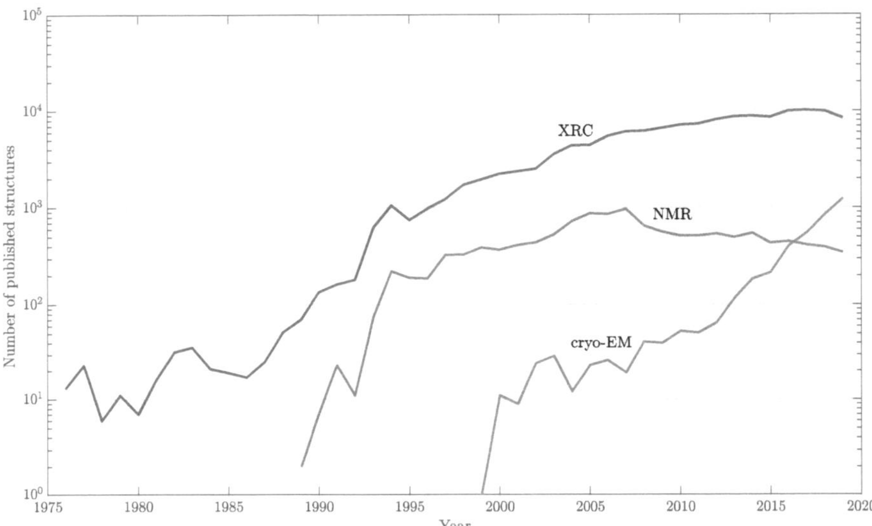

Figure 1.3: Number of published protein complexes in RCSB PDB Here, the number of published structures in each year with respect to its *structural biology* imaging method is presented. Most structural maps result from XRC. Using Cryo-EM the number of deposited maps of protein complexes is growing. NMR spectroscopys main research area is proteins. RCSB PDB statistic as from 19.10.2019

Most protein and protein complex structures result from XRC (see Figure 1.3). This results from the fact that XRC is one of the most established methods. However, cryo-EM started to quickly advance, especially from 2009/2010. The improved hardware, e.g. direct detectors, and further developed image processing software, e.g. maximum-likelihood approach, used for cryo-EM gave the opportunity to reconstruct higher resolved protein complexes [18]. As a result, the popularity of cryo-EM increased. In Figure 1.3 it can be seen that more protein complex structures are published using cryo-EM. There exists a linear upwards trend due to the ability to image small proteins as well as protein complexes. About one tenth of the published protein complex structures in Figure 1.3 result from cryo-EM data. Due to size-limitations of NMR spectroscopy most structures resolved are smaller proteins. The number of possible research objects is limited for NMR spectroscopy. Over the past years cryo-EM has overcome NMR spectroscopy in regard to the number of published structures due to a variety of reasons such as size.

1.2   Single particle cryo-EM is changing structural biology

The aim of SPA cryo-EM is to resolve 3D protein complex maps up to atomic resolution (see 1.4). Hereby, single particle analysis of cryo-EM data means that the reconstructed 3D structure is the back-projection of thousands of averaged recorded single particle projection images. One advantage of single particle cryo-EM is the opportunity to study a variety of different macromolecules. The molar mass of a protein complex can range from 0.1 MDa to 100 MDa. Additionally, it is possible to study symmetrical protein complexes such as the T20S in Figure 1.1a or asymmetric protein-RNA-complexes such as the ribosome in Figure 1.1b. Moreover, the study of different conformations of one protein complex can be done with cryo-EM. The 3D cryo-EM map of the protein complex is a structure which contains the information of the electrostatic potential of the atoms [19]. Another advantage of cryo-EM is that the recorded cryo-EM image encounters the information of the imaged phases [19].

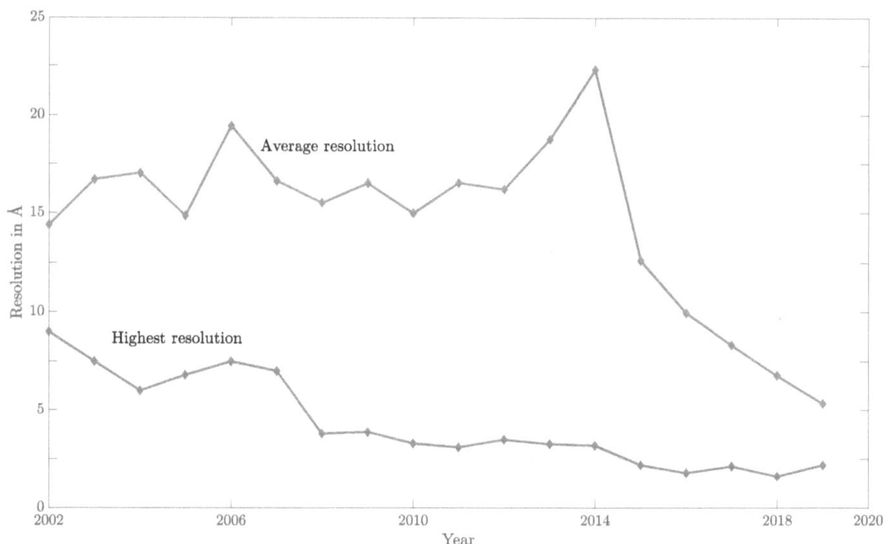

Figure 1.4: Distribution of the resolution of published density maps The two graphs describe the resolutions of single particle reconstructed maps published in the Electron Microscopy Data Bank (EMDB). The brown curve represents the highest resolution of a map published that year. The green graph determines the average resolution of all structures in the EMDB in that year. The data was taken on the 21.03.2019.

With improving components of the TEM and SPA algorithms the chances to reconstruct a greater variety of proteins and protein complexes as well as reaching higher resolutions of these increases (see Figure 1.3 and Figure 1.4). Back in 2002 the highest-resolution of a published structure was less than 10 Å in the EMDB. Six years later, in 2008, the first structures with resolved side-chains were published [20]. Bulky side chains start from resolutions of 4 Å and higher. Furthermore, in 2014 the resolution revolution [21] was a consequence of great progresses in hardware such as new detectors [9] and software. Direct detectors, which directly transmit the signal of the electron to the digital image [20], improved the SNR of the recorded images. As a consequence the increased SNR of the recorded data affected the accuracy of the processing algorithms so that reconstructed structures were able to refine to higher resolutions. Kuhlbrandt [21] went as far as saying that a new era of molecular biology begins. Other influences were the technology to maintain a high qualitative vacuum or maximum-likelihood approach [22]. Furthermore, Nature Methods chose cryo-EM as the "Method of the Year" in 2015 [9] and in 2017 three scientists Jacques Dubochet, Joachim Frank and Richard Henderson were recognized with the Nobel Prize of Chemistry for the work they had done in cryo-EM. Cressey & Callaway [23] cited the Royal Swedish Academy of Sciences stating that cryo-EM has "moved biochemistry into a new ERA". Currently, the highest resolved cryo-EM map is the published structure of the Apoferritin with 1.65 Å (see Figure 1.4) [24].

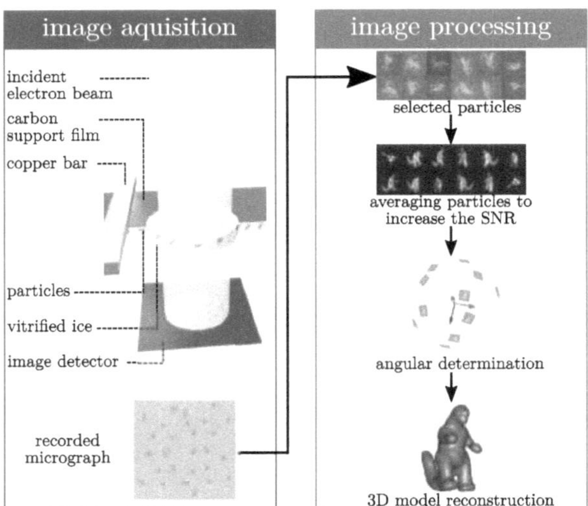

Figure 1.5: General workflow of cryo-EM Here, a general overview over the imaging acquisition on a TEM and the following processing of the data is given. Specimens are either in negative stained or in cryogenic condition. This figure is taken with the courtesy of Wen-ti Lu and adapted.

In general, an incident electron beam as in Figure 1.5 is generated and passes through a specimen in a TEM. On the specimen plane electrons either pass through the sample or scatter due to an object in the specimen. The electron beam is detected by an image detector. The image of the TEM is called micrograph (see Figure 1.5). A single micrograph contains hundreds of single particles. In digital image processing these single particles are identified, cut-out and further processed by algorithms of SPA cryo-EM. The aim of SPA is to optimize a 3D model of the imaged protein complex. The six degrees of freedom of the projection image are maximized with image processing tools such as *REgularized LIkelihood OptiminzatioN (RELION) 3.0* [24] or the *CowSuite* [25–28]. In order to reach high resolution, the raw projection images are iteratively aligned, classified and reconstructed. This process is often called refinement in SPA cryo-EM.

1.2.1 An ill-posed reconstruction problem

During sample preparation a single protein complex is capable to move freely within the liquid solution. The solution is applied onto a grid and rapidly frozen such that the aqueous liquid is immediately vitrified. As a consequence, each single particle is captured in its current orientation in respect to the coordinate systems in Figure 1.6. Every protein complex has six degrees of freedom describing its position within the solid ice layer (see (α, β, γ) and (x, y, z) in Figure 1.6). In Figure 1.6 the synthetic model illustrates the randomly distributed particle within the solid layer on a grid. Indeed, the optimization problem is reduced to a problem of five degrees of freedom due to the projection of the specimen along the z-axis.

In general, a forward model of the relation between the projection image and the 3D map is characterized by $y = Ax$. Here, the variable y is defined by the 2D projection image with respect to the transformation matrix A. In SPA, the matrix A describes the five orientation parameters for a single particle, i.e. the three rotation angles (α, β, γ) (see 2.2.2) and the two shifts in x, y within the specimen. The objective of single particle analysis cryo-EM is to optimize A, the unknown five degrees of freedom of the protein complex with respect to the corresponding projection image. By identifying these parameters it is possible to back-project each recorded projection image. Adding up all these back-projected images a 3D density map of the protein complex is reconstructed. In practice, the model and the transformation matrix are unknown. The maximization step of the orientation parameter becomes an inverse problem $x = A^{-1}y$ [29].

The optimization problem is ill-posed and non-convex. Ill-posed means that reconstructing single particle cryo-EM data most likely misses angular information about the protein complex. Even though a variety of different orientations of a protein complex are given, not all possible orientations of the particular 3D structure are present. The protein complex like the T20S proteasome in Figure 1.1a has orientations which are more preferred than others. The angular distribution over the sphere must not be continuously as it is discrete data. Through sample preparation the homogeneity of the protein complex data is

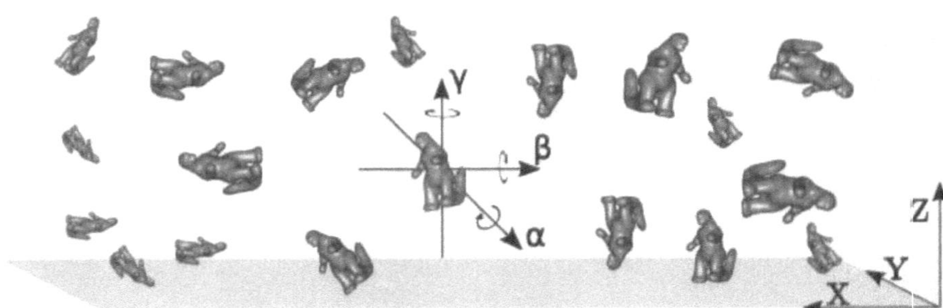

Figure 1.6: Ill-posed reconstruction problem in cryo-EM Here, a cryo-EM grid is sketched. The synthetic complex, the dinosaur, represents a protein complex. Hundreds of particles are present on a single grid. All particles are randomly distributed within the ice. Each particle is rotational shifted by (α, β, γ) with respect to another particle. After the particle identification on the micrograph the regions are cut-out and need to be translated to the center of the projection image.

increased. However, due to radiation damage or heterogeneity based on the conformations of the protein, the optimized structure is only one optimal representation of the protein complex. Different refinement runs as well as different image acquisitions of the identical protein complex can lead to smaller changes in the protein complex structure, which in turn is a second locally maximal optimized structure. The consequence is that there does not exist a global maximum of the protein structure. The reconstruction is not convex.

1.3 Challenges of reconstructing single particle cryoEM data

The optimization problem has two main drawbacks with respect to cryo-EM data. One of these two problems is related to the high noise power. The other one results from imaging a protein complex, which is a WPO, with the TEM.

In general, the electrons, which are scattered by the protein complexes, undergo a phase shift. The projection images recorded with the TEM incorporate these shifts as the phase contrast. However, the protein complex is a WPO (see subsection 2.3.2). It means that the complex is too small to introduce a phase shift that generates a sufficient phase contrast in the recorded image. Therefore, to visualize the single particles on the grid an additional phase shift is constructed by defocusing the objective lens. The additional introduced phase shift needs to be removed for the reconstruction of the cryo-EM data. There is no perfect microscope. Lenses have similar optical defects as in a light microscope. Astigmatism or spherical aberration are some of the perceived effects. The alignment of the TEM determines the quality of the incident beam. If the lenses and apertures of the TEM are aligned well, the aberrations, e.g. astigmatism, which affect the image quality, are minimized. The recorded micrographs contain spread signal information. The CTF correction is the image processing step, where the single particle projection images are correct for these aberrations (see 1.3.1). Hence, a miscorrection of the cryo-EM data can lead to an erroneous refinement of the 3D protein complex structure.

The noise is the unpredictable disturbance of an ideal (resp. predicted) signal (see subsection 1.3.2). Cryo-EM data is very noisy. This results from the fact that biological samples are radiation sensitive. Radiation sensitivity means that the electrons, which are inelastically scattered (see subsection 2.3.1), interact with the protein complex so that the protein complex most likely changes its structure. As a consequence, the electron dose used for image acquisition needs to be kept to a minimum. However, the low electron dose leads to a poor SNR of the raw single particle images, which means that the power of the noise is overshadowing the power of the protein complex signal. The noise further influences the optimization of the recorded cryo-EM projection images (see subsection 1.4.1).

1.3.1 Contrast transfer function (CTF)

Optical aberrations in a TEM introduce blurring to the images. Electron dose or spherical aberration, e.g., affect the detected signal. The introduced phase shift by defocusing the TEM to force a better phase contrast in the images needs to be removed. The Point Spread Function (PSF) corrects for these kinds of defects. The function describes the ideal mapping of a point source in the object onto the image for an optical system. To correct the projection image with PSF the image is convoluted with the PSF. As mentioned in Theorem 2.2.5 the convolution in real space is the equivalent of a multiplication in the Fourier domain. Since the convolution over 2D projection images is a time-consuming calculation the projection images are Fourier transformed (see Theorem 2.2.3) and multiplied by CTF, the Fourier space equivalent of the PSF. The CTF, see Figure 1.7, is an oscillatory, sinusoidal function of spatial frequencies. A multiplication with the CTF Equation 1.1 corrects the displaced phases of the Fourier transformed image. Interpreting any single particle projection image beyond first zero crossing of the CTF is not possible if the CTF correction is skipped [30].

$$CTF(s) = \sin\left[2\pi\left(\frac{\lambda^3 s^4 C_s}{4} - \frac{\lambda s^2 \delta f(\theta)}{2}\right)\right], \tag{1.1}$$

where $s = \sqrt{s_x^2 + s_y^2}$ is the length of the two-dimensional spatial frequency vector and θ is the phase with respect to the spatial frequencies. The wavelength λ (see subsubsection 2.3.1.1) depends on the electrons accelerating voltage used for imaging. The CTF describes the

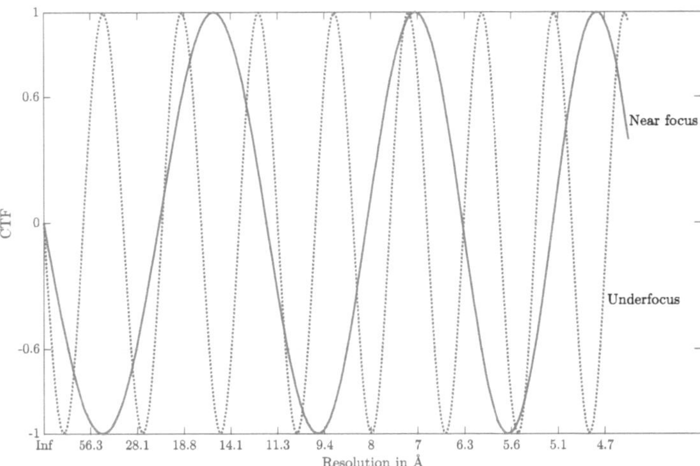

Figure 1.7: Synthetic CTF Here, two CTFs are sketched. The CTF with the near focus is a slower varying sinusoidal function. Here, the defocus is set to $\Delta f = 0.25\ \mu m$, which is close to the back focal plane. The CTF with underfocus corresponds to an imaging with higher defocus. The CTF is varying much faster. Both sinusoidal functions are plotted with the same parameter setting. Parameter: $C_s = 2.7\ mm$, pixel per Å $= 1$ Å, $\lambda = 0.0197$ Å

introduced *defocus* δf set for the objective lens of the TEM. A focused image exists when the beam converges on the back-focal plane. Underfocus and overfocus converge either above or below the back-focal plane. In Figure 1.7 two CTFs with different defocus settings are plotted. With increasing defocus the wavelength of the sine waves decreases. The *spherical aberration* of a lens, called C_s, in the TEM is a constant value with respect to the microscope. It is the inability of the lens to converge the beam to a single focal point at high angles. The resulting image is blurred. Using cryo prepped data the TEM settings are set to underfocus to enhance the contrast of the projection images. All three parameters δf, λ and C_s are known by microscope settings.

Other defects of the TEM, e.g. astigmatism, change the defocus settings of the microscope. Astigmatism leads to different foci with respect to perpendicular rays. It results from either lenses with a non-uniform electromagnetic field [31] or not perfectly centered aperture. Additionally, astigmatisms can occur from beam deflection due to charges from dirty apertures. It creates elliptic shaped Thon rings in Figure 1.8b in micrograph power spectrum. The astigmatism results in a deviation of the defocus based on the phase values. The new defocus values δf_{ast} are determined by fitting the rings of the CTF to the Thon rings, i.e. rings in the power spectrum, of the micrograph. The defocus δf in Equation 1.1

is altered to

$$\delta f_{ast}(\theta) = \delta f_u \cos^2(\theta - \theta_{ast}) + \delta f_v \sin^2(\theta - \theta_{ast}), \tag{1.2}$$

where δf_u, δf_v define the defocus induced along the minimal and maximal axis with respect to the elliptic shaped rings in the power spectrum (see Figure 1.8b). The variable θ_{ast} is the angle between the longest diameter of the ellipse and the Cartesian system with respect to the axis along defocus representation δf_u [30, 32].

There are additional factors, e.g. amplitude contrast, which can further influence the image quality. The envelope function is introduced due to the spatial and temporal coherence of the beam. This function dampens the CTF, especially in the high frequencies. Possible damping functions rely on the drift of the energy spread in the beam or the instability of the current in a lens [30]. A state-of-the-art envelope function is based on the B-factor. Further details are introduced by Mallick *et al.* [30] and Zhang [32].

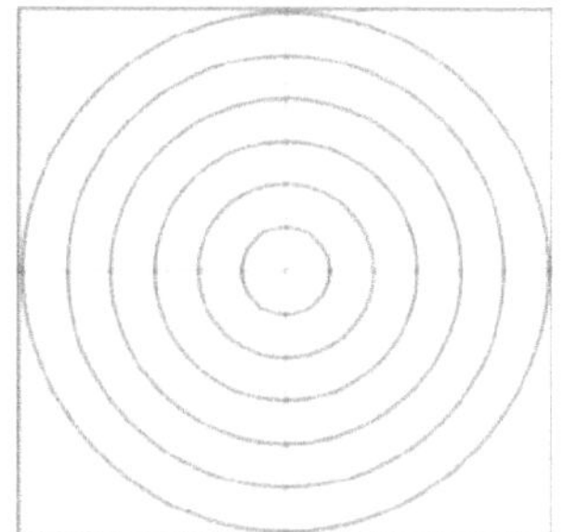

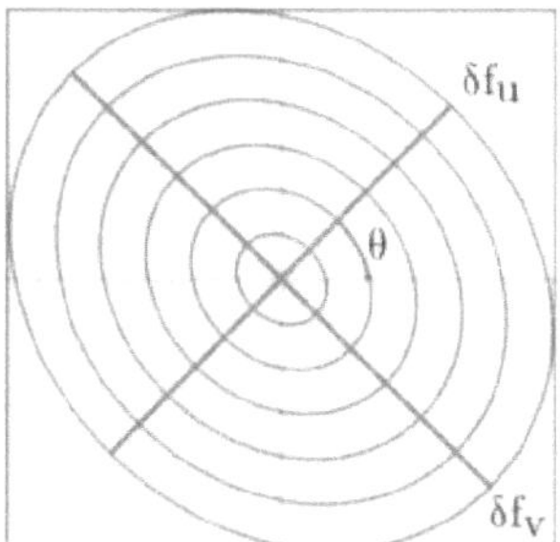

(a) Sketch of a 2D power spectrum with no astigmatism. The CTF is fitted to the power spectrum.

(b) Sketch of a 2D power spectrum with an astigmatism. The CTF is fitted to correct the astigmatism.

Figure 1.8: Correction of astigmatism The teal rings correspond to the maximum peaks of the power spectrum of a micrograph. The CTF is fitted to these Thon rings.

1.3.2 Noise

The objective of an experiment is to measure a particular signal of interest and further analyze and interpret this. The ideal signal in Figure 1.9 is the projection of the synthetic model. Here, the black parts of the image represent areas, where no signal was detected, and the other parts correspond to pixels, where a signal was generated by the 3D density model. In theory, this signal is considered to be the ideal or predicted signal. An ideal signal in cryo-EM is the projection of a protein complex formed by the electron signal. By the resolving power of the TEM the protein complexes can theoretically reach structures with atomic resolution. However, the average published resolution is not reaching the theoretical potential of the method, the atomic resolution (see Figure 1.4). One difficulty is a random process disturbing the ideal projection signal.

Figure 1.9: Synthetic additive image noise The first summand is a projection image of a simulated 3D density map. The map was noise free so that the projection image contains the predicted (resp. ideal) signal. The second summand is a pure Gaussian distributed noise image simulated in MATLAB. The sum of both images represent the measured signal. It is distorted due to a variety of effects.

On the experimental side the measured signal deviates from the predicted signal. A variety of disturbances interfere with the signal of interest. All these combined disturbances are called noise. The noise leads to artifacts, unrealistic edges or blurs out information [33]. Informative content of the noisy image in Figure 1.9 is reduced compared to the ideal projection. Most likely the interpretation of the data based on the measured signal is difficult and leads to false assumption of the underlying structure. In digital image processing noise emerges from image acquisition, image coding, transmission and processing the data [33]. The contamination of a specimen can lead to a false signal. A faulty memory location, e.g., can corrupt the digital image [33]. All these interferences add up to generate the noisy measured signal in Figure 1.9.

In general, disturbances are unpredictable, random and describe the combination of all physical components which interfered with the predicted signal. The characteristics of noise are modeled by probability distributions describing the random statistical processes. The most common distribution of noise is the Gaussian (see in Figure 1.9). There are also Poisson noise, uniform noise and impulse noise [33]. The noise in signal processing is often considered to be a white or colored noise. The power spectrum of the noise defines the color. White noise is image noise, which is normally distributed with zero-mean and variance of one. It has a constant power spectrum with respect to the identical length of spatial frequencies intervals. Colored, e.g. pink or blue, noise occurs with different spectral properties than white noise. Modeling the noise component in image processing is done in two different ways. On the one hand there is multiplicative noise, which depends on the signal. This type is more severe since it is not easily separated from the ideal signal. On the other hand there exists additive noise as in Figure 1.9. The noise is added on top of the signal and does not modify the predicted signal. In image processing theory of SPA of cryo-EM data the random processes are formed as an additive model. A simple representation of a single particle projection image is

$$I = f + m \qquad m \sim \mathcal{N}(\mu, \sigma^2), \tag{1.3}$$

where the Noise m is Gaussian distributed with mean μ and variance of σ^2. In Figure 1.9 the Gaussian noise image was added onto the ideal projection image f leading to a modified image I. This is similar to a single particle projection image where a noisy component was added on the underlying ideal signal.

To define the information value of an image a ratio between the power of the signal and the power of the noise is determined. This ratio is called the SNR. An SNR equal to one indicates the same amount of signal as noise present in the data.

$$SNR = \frac{P_{Signal}}{P_{Noise}} \tag{1.4}$$

The three images in Table 1.1 represent the identical underlying signal but different powers of noise. The first image has about the same amount of power for noise and signal. Here, the SNR is close to one. The other two images contain a greater amount of the additive noise. The second image with an SNR of about 0.25 has about four times more noise power than signal. The signal for the dinosaur tail has lost some visibility. In the third image the tail is completely invisible. Identifying the signal in the images, which is often one aim of image processing, is difficult. A low SNR affects the quality of the image processing results. Therefore, a sufficiently large SNR is necessary to be able to differentiate between the signal and the noise and consequently, be able to correctly extract the signal information. The SNR of cryo-EM projection images is very small. It often ranges from 0.1 to 0.3. To increase the SNR the number of electrons used for image acquisition could be increased, but the radiation sensitivity of biological matter makes it difficult to take images at higher electron dose. As a consequence, increasing the electron dose damages the structure of the protein complex.

The computational techniques aim to remove additive noise in recorded data depend on the noise sources. Noise is often caused by multiple aspects during image acquisition. The model describes all sources that caused the random disturbances of the signal. The quantization error, e.g., emerges from the transmission of a continuous signal to a measured digital discrete signal [33]. In general, any wave function is in theory a continuous function. The signal generated by electrons can only be measured at finitely many time points. Therefore, there exists a difference in the ideal signal to be detected and the discrete on the spatial-scale depending signal. The mapping of the spatial frequencies to a pixel is not precise and furthermore, deducts the signal information quality. Quantization error is often assumed to be additive white noise. Thus, it is important to learn and understand the noise

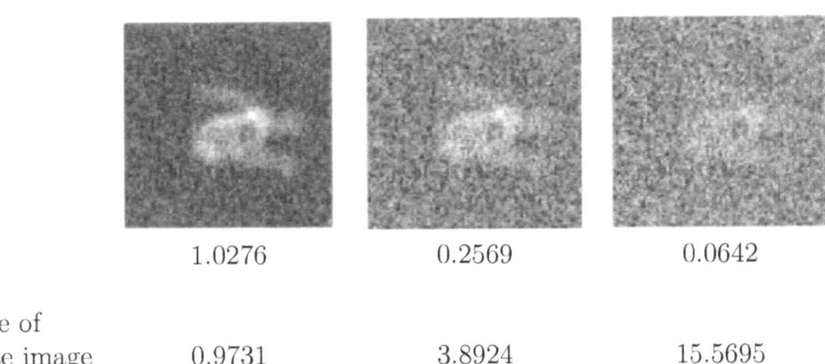

SNR	1.0276	0.2569	0.0642
Variance of the noise image	0.9731	3.8924	15.5695

Table 1.1: SNR of synthetic data Here, three projection images of a synthetic map with different SNR values are presented. All three of them contain the identical power of the signal. They differ in their power of additive component of noise σ^2. With decreasing SNR the signal of the maps features are more invisible. The tail of the dinosaur is a finer detail of the synthetic 3D model. The additive noise power covers the power of the signal with respect to this particular feature.

source before going into image processing.

1.3.3 Noise in cryoEM data

The electrons in a TEM are scattered by the protein complexes. In the best case the ideal electron signal is detected and digitized. In real world applications the signal is disturbed due to, e.g., the physical behavior of the electrons. Baxter *et al.* [34] categorized noise occurring in the TEM into *shot noise*, *structural noise* and *digitization noise*. Besides these, the scattering interference resulting from the nature of electron scattering can be seen as noise. Furthermore, the concept of salt and pepper noise deals with corrupted image pixel values. This is related, e.g., to defect pixels on the detector. Often these corrupted pixels are set to a specific value such as the maximal value or mean value of the other pixels. Optionally, these hot pixels are set to zero. In Figure 1.10, an exemplary micrograph with pure noise related information is shown. It does not contain a protein complex signal. However, it shows the variation of the noise. The power spectrum of the micrograph shows low spatial frequencies. Even though noise is a random process, it generates a signal, which interferes with the protein complex signal. The main noise sources in the TEM are explained in more detail.

Scattering   The scattering of an electron is not always elastic forward scattering. Some electrons are back-scattered, others are scattered multiple times or inelastically. Scattering is also related to the sample thickness. With increasing thickness of a specimen more than one scattering process, i.e. multiple scattering, is enhanced. Here, the scattering angle detected is a combination of scattering angles leading to a signal which is complicated to

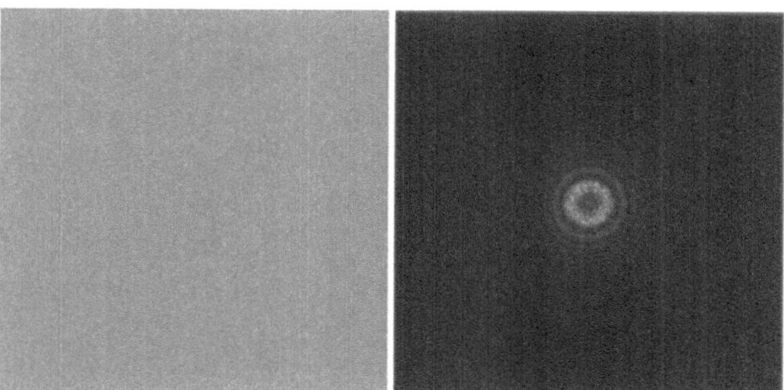

Figure 1.10: Noise micrograph and its power spectrum Here, a micrograph recorded with a TEM is shown. The micrograph is the result of imaging a grid with a thin carbon support film. On the right side the corresponding power spectrum of the micrograph is presented. The power spectrum was computed with *CowSuite* [25–28].

interpret. Another unwanted scattering type of electrons is the inelastic scattering. These contribute to the noise component of the recorded micrograph. Their energy loss causes, e.g., beam damage, secondary electrons or X-rays. The scattering interference with the predicted signal of different single particle projection images is independent.

Structural noise Structural noise is related to any electron being deflected by an atom which is not part of the intact protein complex. Exposing a biological specimen to electrons leads to an interaction. Especially, inelastic scattering, i.e. electrons which undergo a change in energy, affects the stability of the protein complex structure. Due to inelastic scattering, electrons in the protein complex could leave an orbit, which leads to an ionization of the particle. As a consequence, the structure of the protein complex is harmed. This is called radiation damage. If a protein complex is broken in the specimen, it results in false signal compared to the ideal predicted signal of that protein complex. [34]

Secondary electrons are electrons, which were kicked out of the atoms' electron orbit. These can again be deflected by a protein complex producing a signal in the projection image which cannot clearly be traced back to the ideal scattering in the specimen [34]. A thin carbon support film can also add to the noise. In addition, if the ice is not perfectly vitrified during plunge-freezing, there are ice crystals in the sample. These ice crystals also deflect electrons which in turn results in a disturbance of the ideal signal. All electrons scattered by ice crystals in the specimen interfere with the signal of the protein complex. This signal is not homogeneous over the whole micrograph. [34]

Shot noise *Shot noise* results from the natural behavior of an electron. The current, which produces the electromagnetic field of the lens in the TEM, is not consistent throughout the lens. Indeed, it has a number of different discrete charges. The electrons pass through the electromagnetic fields and have to overcome the potential barriers. As a consequence, there are statistical fluctuations. Shot noise is independent of other electrons. Shot noise is Poisson distributed [33, 35]. The Poisson distribution is based on a fixed number of events occurring in a specific time interval. All events are time independent and appear with a specific constant mean value. Computational algorithms based on the Poisson distribution are far more complicated. Therefore, the noise component is modeled by the Gaussian distribution since the Poisson distribution converges to the Gaussian distribution for large observation numbers. [33]

Detector noise The third stage of adding noise occurs while detecting and reading out the signal. The detector noise is related to the nature of radiation, detector material and spatial frequency. In most TEMs the electron wave is being recorded by a direct detector 2.3.1, which is transfered and digitized into an image. The incoming signal is a continuous function, which is digitized into a discrete function. This noise component is considered to be of Gaussian nature. The detective quantum efficiency (DQE) is the ratio between the input SNR and the output SNR [36]. It describes the efficiency of the direct detector detecting the electron signal and transforming it into images.

All these effects lead to a disturbance of the ideal protein complex signal. Each projection image includes a specific combination of these noise components. Thereby, the noise is not always distinct. During image acquisition (see section 2.3) the specimen can be recorded multiple times. The resulting micrographs have the same shot and background structure noise but a different digitization noise [34]. Hence, it is important to understand the noise formation during imaging and processing the protein complexes.

The ideal signal of the protein complex is unknown and the power of the noise particularly high compared to the signal. To quantize the noise in the readout images is challenging. The noise is statistically modeled. In cryo-EM data, the noise is assumed to be Gaussian distributed with the properties of zero-mean and variance of one. Thus, all projection images are assumed to encounter the same underlying Gaussian distribution for the noise. Hence, one aim of cryo-EM image processing is to reduce the noise by averaging projection image (see section 2.4).

1.4 Resolution

The aim of SPA is to visualize single atoms in the reconstructed protein complex map. The resolution of a 3D map defines a point up to which specific resolved features are present. With increasing resolution the structure of the protein complex is more detailed so that the interpretation of the function of the protein complex is more in depth. The atomic resolution of a protein complex as a construct of atoms is based on the visibility of single atoms, e.g. hydrogen with diameter 0.74 Å.

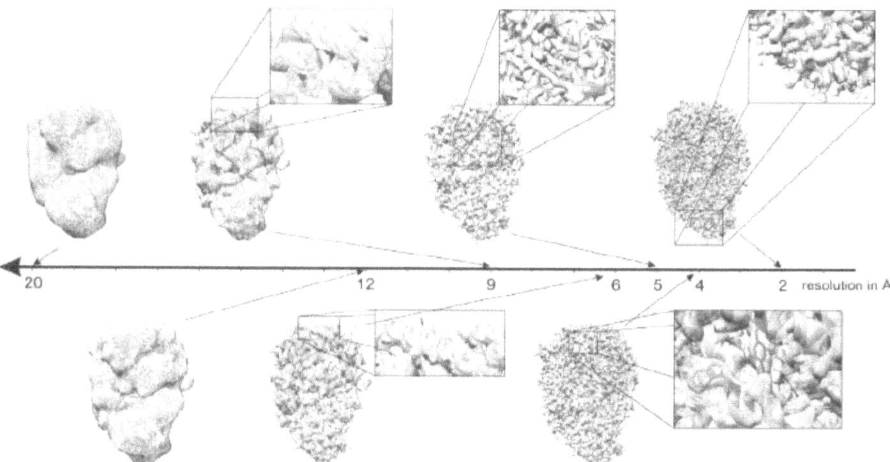

Figure 1.11: Spatial resolution of protein complexes A protein complex has certain features that occur with a certain frequency resolution. At a low resolution of 20Å the protein complex seems to be a smooth volume. Resolutions around 12Å to 9Å show larger and smaller regions and define e.g. subunits. Starting from 3 Å chemical features such as side chains are resolved. Almost all single atoms appear from a resolution of 1 Å. Dr. David Haselbach provided an overview of features using the atomic model of a CRM1-Ran(GTP)-snurportin complex (pdb: 3gjx). The figure is used with the courtesy of Dr. David Haselbach.

Features in protein complex, in general, are based on chemical properties of the complex (see Figure 1.11). In Figure 1.11 it is easy to see that a protein complex is more or less a smooth surfaced object when it has a resolution of around 20 Å. A resolution lower than 10 Å gives only a rough estimate of the domains in a protein complex. It is not possible to distinguish atoms or even see amino acid side chains. From 7 Å the alpha helical becomes visible. A reconstructed map below 4 Å has bulky side chains visible. Further, with increasing resolution more details of the complex such as β-sheets or side-chains become visible. Structures below 2 Å show atomic features such as water molecules and ions. Around 1 Å almost all atoms of a protein complex should be visible in the refined map. Mathematically, the feature resolution refers to either the point resolution, where a point marks the smallest resolved feature, or the sine resolution, which is based on the highest spatial frequency present in the data [37]. Penczek [38] defined resolution of a 3D map as the shortest distance between two distinguishable features in the sample. It is impossible to have a higher sine resolution of a refined map than the resolving power of the instrument with which the projection images were taken [37].

1.4.1 Influencing factors of the feature resolution

Even though, in theory, the TEM can resolve objects that are in size smaller than the diameter of an atom, 3D cryo-EM maps are not always resolved to high resolutions. The feature resolution of the reconstructed protein complex is affected by the biological behavior of the complex, the image acquisition and the image processing tools.

1.4.1.1 Number of projections

The resolution of 3D protein complex structures is affected by the number of distinct projection images used for the reconstruction. A cryo-EM data set with a variety of different projection angles present has a better representation of the 3D rotation group, which contains all possible rotations about the origin of the 3D Euclidean space. During Fourier reconstruction the cryo-EM projection images overlap in their central sections (see Theorem 2.2.6). The amount of overlap depends on the number of projections and the dimension D of the protein complex. The consequence is that the feature resolution of cryo-EM data is limited by the number of projections N and their angular distance $\Delta\phi = \pi/N$. The maximal theoretical feature resolution g is defined by the following equation

$$g = D \cdot \sin\left(\frac{\Delta\phi}{2}\right) \quad = \quad D \cdot \sin\left(\frac{\pi}{2N}\right) \quad \underset{\text{for large} N}{=} \quad \frac{\pi D}{2N}. \tag{1.5}$$

The maximal resolution, which can be achieved by three projection images, is $1/2$ of the dimension D of the protein complex [37]. Nowadays, it does not influence resolution of the refined cryo-EM data. The advancements in hardware made it possible to record enough data.

1.4.1.2 Nyquist Shannon Sampling Theorem

For high resolution structures it is necessary to detect the complete signal related to the protein complex. In order to digitize the signal it needs to be sampled from a continuous into a discrete signal. A sufficient sampling frequency f_s is necessary to transfer the detected signal, an electron wave, to discrete points without the loss of signal information [39, Ch.4]. The Nyquist Shannon Sampling theorem derives the minimal sampling frequency f_s, called *Nyquist frequency*, to adequately convert a continuous signal into a discrete digital image in the TEM. Let W be the maximum frequency of the signal of interest. The sampling frequency f_s has to be twice the maximum signal frequency

$$f_s \geq 2 \times W, \tag{1.6}$$

so that the signal is stored without any information loss. If this relationship is considered during image acquisition, the resolution of the protein complex should not be affected.

1.4.1.3   Noise influence on image processing

The noise (see subsection 1.3.3) influences the SPA (see section 2.4), which has an impact on the feature resolution of the reconstructed map. It depends on the accurateness of the stored signal. Due to various effects such as aberrations of the TEM or noise (see in subsection 1.3.3) the protein complex is not resolved up to atomic resolution. Moreover, the noise can cause a misinterpretation of the observed data and more importantly false reconstructed maps. The three primarily noise influencing parts are called *shot noise*, *structural noise* and *digitization noise* (see subsection 1.3.3). In general, the noise in the data is a combination of those effects and therefore, often difficult to quantify. The aim of image processing is to reduce the noise within the projection images and therefore, enhance the SNR. However, the variation of the noise affects the image processing tools. A reliable alignment, e.g., of the protein complex with these SNR values is often impossible. Additionally, parameters, provided by the user, within the refinement algorithms are often specific to the protein complex. The masking parameter for the recorded projection images, e.g., depends on the diameter of the protein complex which in turn is specified by the user. A tight mask around the protein complex cuts off the protein complex's signal. A mask with a significantly larger diameter than the protein complex takes too much noise information into account. The identical mask is applied to both half-sets. Consequently, there exists a well-correlated part within the two half maps. Multiple effects such as model bias influence the refinement of the cryo-EM data such that the structure of the protein complex is not the ideal representation.

Overfitting noise   The *OXFORD* [40] states that overfitting is "The production of an analysis which corresponds too closely or exactly to a particular set of data, and may therefore fail to fit additional data or predict future observations reliably." In general, overfitting has a low bias and a high variance. To overfit noise means that algorithms optimize until the noise becomes part of the ideal signal. The noise in cryo-EM is assumed to be uncorrelated Gaussian noise. The variation of the noise in the projection images can bias the identifying, aligning or classification of the cryo-EM data. If algorithms tend to compute too detailed information, then the noise could fit the variation of the optimized system and hence, be detected. This leads to a correlation of the noise components of different single particle images.

Model bias   Model bias is one of the main issues related to cryo-EM data. Sigworth [41] stated that model bias, in general, is the impact of a reference to the reconstruction. This effect is not particular to an image processing step but results throughout any reference related computation. Template picking and projection matching are typical examples for pushing the cryo-EM data towards a specific appearance of the object.

Identifying single particles   Working with biological matter has its limitations. Sample quality is essential to determine high-resolution structures. Samples contain a layer of ice and carbon support film. The unspecified thicknesses of the ice and the support film affect the contrast of the projection images [42]. The lower the image contrast the more difficult it is to identify the single particles on a micrograph. The issue is to ensure that picking algorithms (see section 2.4) detect signal which is related to particles instead of noise. Besides, projection artifacts can be found due to the lack of depth sensitivity. The cryo-EM projection images are generated through transmission over the 3D protein complexes in a TEM. It can lead to false assumptions of the imaged structure [31, Ch.1]. Furthermore, micrographs containing thousands of particles close to each other push particle picking/selection to their limits.

Classifying different conformations   Specimen heterogeneity is controversial to being an advantage or disadvantage. Cryo-EM has the ability on the one side to capture particles of different conformations in one sample. On the other side too many different conformations on a grid can lead to computational issues resulting in low-resolution structure. As mentioned in section 1.1 protein complexes are dynamic objects and hence, occur in different conformations. If the sample is not sufficiently purified, too many different conformations or other proteins are visible in the specimen. This often leads to an insufficient number of similar protein complex projection images. One computational difficulty is to sort out and refine these data sets. During classification different conformations should be sorted into different sub-classes. However, the noise dominates the higher spatial frequencies such that classification routines may fail to sort the data into distinct and clean classes. As a result the variety of projection images, based on the variability of the detected signal, cannot be averaged to reduce the noise, which causes a poorly improved SNR. This low SNR implies a poor performance of reconstruction tools. It leads to low resolved 3D protein complex structures. Hence, it is important to minimize the conformation variation in one dataset. Sigworth [41] experimented identifying and grouping heterogeneous samples with respect to a decreasing SNR.

Often the protein complexes have rigid and dynamic regions. As a consequence these dynamical parts underly the similar classification issues and result in low-resolution structures. Radiation damage of the specimen leads to the image acquisition of broken structures of the protein complex. These projection images also vary from the ideal signal and underly the classification issues.

In general, the feature resolution of the 3D reconstructed map is complicated to compute. The details as presented in Figure 1.11 are difficult to measure. If the structure is a low-resolution reconstructed map like 10 Å, the details present are too coarse so that pinning a number to the resolution is difficult. Therefore, statistical methods like the FSC are used.

[38]

1.4.2 Fourier Shell Correlation (FSC)

The FSC is the correlation between two Fourier transformed volumes (for details about Fast Fourier Transform (FFT) see subsection 2.2.3). The spectral consistency of the two 3D maps in Fourier space is evaluated. Therefore, it divides each Fourier map into shells and correlates the two Fourier volumes in respect to these. Hereby, it takes the amplitude (2.14) as well as phase (2.15) information of the two objects into account. The 2D equivalent is called Fourier ring correlation.

Theorem 1.4.1 (Fourier Shell Correlation [38]) *Let U and V be two 3D reconstructed volumes in Fourier space, then the FSC is defined as*

$$FSC(u, v; s) = \frac{\sum\limits_{||s_k|-s|\leqslant\epsilon}^{k_s} U(s_k)\overline{V(s_k)}}{\sqrt{\left(\sum\limits_{||s_k|-s|\leqslant\epsilon}^{k_s} |U(s_k)|^2\right)\left(\sum\limits_{||s_k|-s|\leqslant\epsilon}^{k_s} |V(s_k)|^2\right)}}, \tag{1.7}$$

where $\overline{V}$ denotes the conjugate Fourier transform of $V(s_k)$ (see (2.17)). s is the 3D spatial frequency vector and s_k is the k-th shell in Fourier space.

An alternative representation of the numerator of the FSC is

$$U(s_k)\overline{V(s_k)} = |U(s_k)|\,|V(s_k)|\cos(\Delta\Phi_{(U,V)}(s_k)), \tag{1.8}$$

which makes it clearer that the FSC depends on the amplitude of the two maps and their phase difference [38]. The denominator and numerator in the FSC are similarly defined as the normalized cross-correlation (NCC) (see subsection 2.2.1). The difference, here, is the sum over all voxels in the 3D map compared to the shell-wise summation. It can be seen as applying a band-pass filter (see section 2.4.1) to the maps. The currently summed shells in each of the two Fourier volumes are multiplied, whereby all the other shells are set to zero. In Figure 1.12, the FSC curve shows an exemplary characteristic of the correlation between two maps. In general, correlation values, which are equal to one, correspond to a high resemblance between the two compared objects. Values around zero correspond to no similarity in the data. Data is considered to be consistent whenever the phase and amplitude of the Fourier transformed object are similar at the same spatial frequency.

The FSC has been established as the state-of-the-art measurement for the resolution of cryo-EM data. The FSC is supposed to determine the point until which well resolved features of the protein complexes as presented. Therefore, the FSC is used as a resolution

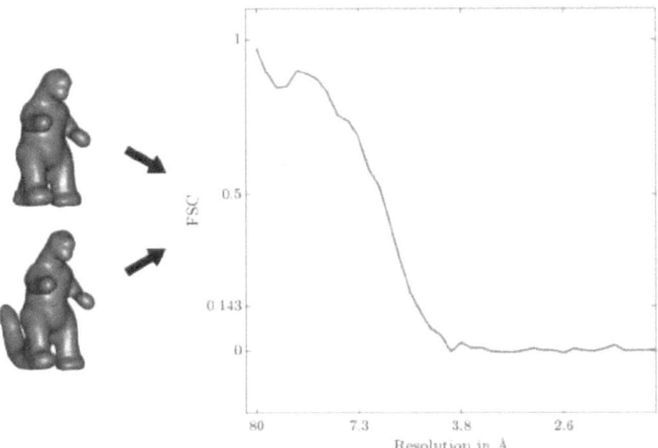

Figure 1.12: Exemplary FSC The two maps to the left side are a synthetic data. One of them is missing a tail. Hence, the correlation between them cannot be equal to one over all frequencies. Both 3D maps were overlaid with an artificial noise by the CowSuite. The pixel size was assumed to be 1 Å per pixel.

criteria in the field of cryo-EM. There are two common processing routines for the cryo-EM data in order to measure the resolution with the FSC. One possibility is to split the data into two half sets and independently align and reconstruct. The other possibility is to do the alignment with the whole cryo-EM data set and afterwards divide the set into two subsets for the reconstruction. The aim of partitioning the data is to refine the sets independently. The reconstructed 3D cryo-EM half maps should ensure a reduced biasing of the data by the noise and input references [38]. This half set processing method was first mentioned by Van Heel [43] and is often referred to as the *gold standard* [22]. [22, 37, 38]

The feature resolution (see section 1.4) is estimated at the point where the FSC drops below a specific value. The two cryo-EM maps are assumed to be well correlated at this threshold. Different levels to cut-off the FSC have been proposed. The more conservative threshold 0.5 was suggested by Böttcher *et al.* [44]. At this point the power of the signal and the power of the noise (resp. for the full or half data set) are equally present. If the cryo-EM data was gold-standard refined, the resolution is often measured at a point, where the FSC drops below 0.143. Rosenthal & Henderson [45] derived this number based on the correlation between the full data reconstruction and a reference map of the protein complex. Hereby, Rosenthal & Henderson [45] links the reference correlation to the FSC between the two half-set reconstructions. In Figure 1.12, the two thresholds are shown. Furthermore, Van Heel & Schatz [46] introduced the $\frac{1}{2}$ bit information level. This threshold criterion has not been established to estimate the resolution in the field of cryo-EM.

1.4.2.1 Drawbacks of the FSC

The nominal resolutions of published cryo-EM maps show a positive trend in reaching higher and higher numbers over the past years (see Figure 1.4). The feature resolution of these maps is measured with the FSC. Furthermore, it has been established that the FSC is used as the resolution criterion for the refined 3D protein complex structures in the field of cryo-EM. The FSC is used as the resolution criterion for cryo-EM density maps. As a statistical measurement, it is subjected to a variety of influences.

Van Heel & Schatz [46, 47] discussed many mathematical issues influencing the quality of the FSC and its interpretation of the resolution of the reconstructed cryo-EM maps. The FSC fails in respect to, e.g., decreasing SNR or masking. If both refined maps are masked such that parts of the 3D map are set to zero, then the Fourier volume is affected and the FSC could overestimate the resolution [48]. The size of the structure within the reconstruction box influences the FSC. The radius of each shell and the number of voxels within these shells affect the quality of the FSC. Further on, the symmetry of a protein complex influences, in fact, the number of independent voxels within a shell. With increasing symmetrical units the number of independent voxels is reduced. The estimated resolution based on the FSC is less reliable. Ideally, correction factors accounting the number of voxels or symmetrical units are applied to the shells. [43, 46].

The threshold, which defines the resolution, is under debate [43, 45, 46]. They are discussed to be too conservative or too optimistic. It is often questioned whether the FSC still reliably measures the reconstructed signal or correlates the noise present in the protein complex maps. Thereby, one of the main concerns is that the assumption of uncorrelated noise in cryo-EM data does not hold. As introduced in this chapter the noise of cryo-EM projection images is white Gaussian distributed with mean zero and variance of one. After processing the data the noise might be subjected to overfitting (see subsubsection 1.4.1.3), which could result in a high correlation. Because of the low SNR of cryo-EM data it is complicated to determine whether the threshold of the FSC is confident or not. Especially, in the higher spatial frequencies the amplitudes of the Fourier transformed images are dominated by the noise [49]. Therefore, a large amplitude in the Fourier volume must not correspond to a strong signal of the protein complex. As a consequence, the FSC could correlate noise. The FSC is sensitive to the noise in the cryo-EM data.

With the advancement of the cryo-EM method the published maps tend to resolve to higher resolutions. However, some of these protein complex maps mismatch in their claimed resolution and qualitative, visual assessment. If the FSC claimed a resolution of 4.5 Å, however, features such as α-helical or β-sheets are not visible, the quality of the 3D cryo-EM map and its resolution should also be questioned. The disagreement of the visible and the estimated resolution could be caused by the noise in the cryo-EM data. As mentioned, the noise influences the image processing (see 1.4.1.3). During the alignment, e.g., the low SNR of the cryo-EM data could lead to a misinterpretation of noise as the recorded signal. Noisy parts of the single particle images are aligned and reconstructed. Consequently, the Fourier volumes could encounter consistent information due to noise. The FSC measures the similarity of these volumes and hence, overestimates the resolution. Therefore, the FSC is an indicator for the resolution. However, it does not define the feature resolution.

It has been stated that gold-standard refinement ensures an independent data processing, so that the FSC is a reliable estimator. An often discussed controversy about three separately published cryo-EM structures of the same protein complex, the trimeric HIV-1 envelope glycoprotein demonstrates that the FSC is not independent on the data processing style. Three published maps show inconsistencies in their structures. However, all three authors claim well-resolved protein complex maps based on the FSC. The authors picked the recorded micrographs with references and aligned these to the reference. Doubts about certain reconstructed protein complex regions arose [50]. Moreover, the three maps contradict in structural features. Henderson [48], Subramaniam [50], and van Heel [51] criticized the alignment procedures which use a reference to identify particles of the protein complex.

In 2012, the first validation task-force meeting for cryo-EM data took place. It was criticized that the estimated resolution of the published maps is immensely optimistic [52]. Recommendations to process data were given during that meeting. The independence of the two refined maps is also essential to evaluate the resolution. In 2017 again, several researchers met for *The CryoEM Structure Map and Model Challenges* to challenge the published cryo-EM maps. The aim of this meeting was to question the resolution and the quality of the currently published single particle reconstructed cryo-EM maps. The meeting was split into two main topics, the map challenge and the model challenge [53, 54]. The result was that cryo-EM maps are being published with claimed FSC estimated resolutions but lack of an underlying ground-truth of the detected signal. Conclusively, the FSC, even though it is used as the resolution criterion, does not define the correctness of the reconstructed protein complex map. Conclusively, the FSC is no validation tool. The standards for validating data as well as publishing data are not clear throughout the field of cryo-EM.

Validation means to examine data in their correctness. If data is validated, the quality of data is determined [55]. Protein complexes are made of chemical bonds. These have, e.g., specific bond lengths and bond angles. In the concept of cryo-EM the validation of protein complex structures should verify these chemical properties. Those features are known and any divergence should indicate quality issues in the protein complex structures. In cryo-EM, atomic modeling is often started at resolution of around 4 Å. At this point bulky side chains are visible in the cryo-EM map. The modeling of atoms into these side chains is only reliable to a certain portability. The consequence is that the atomic model is often influenced by the refined cryo-EM density map. It has also been noticed that some cryo-EM density maps disagree with the atomic model. If the model is fitted into the cryo-EM map, some of the bond lengths are altered. However, the majority of these bond properties should not differ from the theory. It could be concluded that the refined 3D maps do not visualize the true structural behavior of the protein complex. The consequence is that the published cryo-EM map is not valid.

For some protein complex structures, there have been published maps from other imaging techniques (see 1.1.1). These could be used to review the correctness of the cryo-EM map. However, atomic modeling or comparing other protein complex structures is not a validation based on the underlying ground truth, the detected protein complex signal. In general, the validation of cryo-EM data is complicated. The low SNR of the data makes it difficult to relate the reconstructed signal to the recorded signal.

1.5 Aim

Resolution is not precisely defined in cryo-EM. Currently, the cryo-EM maps claim their resolutions based on the FSC. However, the FSC does not provide a quality assessment of the reconstructed chemical features. Furthermore, the noise in cryo-EM data influences the cryo-EM image processing. Various aspects have an impact on the resolution and quality of the cryo-EM maps. Misinterpretations of the recorded signal propagate into false protein complex structures and an overestimation of the reconstructed protein complex maps. The objective of this **book** is to collect and evaluate aspects of the noise influence. Three different experiments are carried out. These experiments provide difficulties with respect to the alignment and classification of the noisy cryo-EM data. Another experiment is related to the image acquisition of WPO. The modification of CTF correction can result in the qualitative loss of the correctness of the reconstructed protein complex structure. These experiments stress the importance of a cautious and clean image processing style.

Furthermore, the aim of this **book** is to provide a feature resolution measurement for reconstructed cryo-EM structures. An approach, which is supposed to validate the true resolution of the refined protein complex structure, should define the quality of reconstructed data based on the experimentally recorded data. Assuming that noise did not affect the image processing steps, the reconstruction projection images should contain a similar power and variation of signal present as in the experimental data. The low SNR of the recorded data makes it difficult to distinguish the true recorded protein complex signal. The variation of the noise often agrees with the variation of the signal for high resolution information. Additionally, the noise is complicated to be quantified. The consequence is that the noisy cryo-EM data cannot be divided into signal and noise. The goal is to define a metric that inspite of the noise measures the distance between the recorded single particle projection image and the corresponding projection of the reconstructed protein complex along the same degrees of freedom.

Outline In Materials and Methods the mathematical concepts such as the Fast Fourier Transform or the central-slice theorem are introduced. Further on, a detailed introduction to image formation and processing of cryo-EM data is given. In Map assessment computational algorithms to measure the spatial resolution of protein complex structures are presented. In Results three noise experiments are analyzed. The focus of Results lies on presenting a validation method and outline the testing of this algorithm. In Discussion an interpretation of the results is presented and Conclusion and Outlook gives an overview of the topics to be investigated.

Chapter 2

Materials and Methods

2.1 Software

Software	Link
Matlab 2017b	de.mathworks.com
Matlab 2018a	de.mathworks.com
CowSuite	www.cow-em.de
Relion 1.3	bitbucket.org/scheres/relion
Relion 2.1	bitbucket.org/scheres/relion
Relion 3.0	bitbucket.org/scheres/relion
Chimera	www.cgl.ucsf.edu
ChimeraX	www.cgl.ucsf.edu
Gautomatch	https://www.mrc-lmb.cam.ac.uk
Inkscape	[56]

Table 2.1: Software

cryo-EM software State-of-the-art software to refine cryo-EM data is RELION developed by Zivanov *et al.* [24]. Gautomatch is a software used to fully automated pick single particle in cryo-EM. Chimera as well as ChimeraX by Goddard *et al.* [57] are capable to display cryo-EM maps and atomic models. The CowSuite is a software tool for single particle analysis in cryo-EM. The suite is developed by the Department of Structural Dynamics of Prof. Holger Stark (Luettich [25], Heisen [26], Busche [27], Kirves [28], Schulte [58], Lambrecht [59]). This collection of software is used to process data from the micrograph all the way to the structure map. There are the CowEyes, the Micrograph Quality Checker (MQC), CowGrace by Schulte [58] and particle picker called John Henry by Busche [27].

MATrix LABoratory MathWorks designed MATrix LABoratory (in short MATLAB) for mathematical computations, especially numerical nature. All routines, functions and scripts in Appendix B are implemented in MATLAB 2017b and MATLAB 2018a. MATLAB has a variety of functionalities in the main framework, but for specific thematic areas additional toolboxes (see Table 2.2) are included.

Toolbox	Version
Image Processing Toolbox	Version 10.2
Communications System Toolbox	Version 6.6

Table 2.2: **MATLAB Toolboxes** These toolboxes are used in this **book**.

2.2 Mathematical Preliminaries

To understand the computational background of cryo-EM image processing, general mathematical definitions such as the Fourier transformation or the projection of a 3D map are necessary. Furthermore, certain terms such as image mean and image variance are commonly used in the field of cryo-EM. Here, the definition of the statistical terms for a 2D image are clarified. Moreover, the FSC is the correlation between two variables in Fourier space. The SSNR, which is introduced in 2.5.1, is also a measurement in Fourier space. The validation approach derived in 3.2 is based on the SSNR and the FSC. As a consequence the Fourier space and its advantages are necessary to be established. Basic mathematical concepts such as image statistic (subsection 2.2.1), projection and rotation (subsection 2.2.2) and *Fourier Analysis* (subsection 2.2.3) are introduced.

Euclidean distance   The euclidean distance measures the length of vector $x \in \mathbb{R}^K$ with $K \in \mathbb{N}$. It can also measure the distance between two points in the vector space.

$$d(x, y) = \sqrt{\sum_{k=1}^{K} |v_k - \xi_n|^2}, \tag{2.1}$$

where the vectors are described by $x = (v_1, \ldots, v_K)$ with $v_k \in \mathbb{R}$ and $y = (\xi_1, \ldots, \xi_K)$ with $\xi_k \in \mathbb{R}$. The euclidean distance is often written as $d(x, y) = \|x - y\|$. The vector spaces $\mathbb{R}^K$ and $\mathbb{C}^K$ are equipped with the euclidean distance.

Gaussian distributed   A random stochastic variable x is Gaussian distributed with mean μ and variance σ^2 if the variable x fits the following distribution function.

$$f(x|\mu, \sigma^2) = \frac{1}{\sqrt{(2\pi\sigma^2)}} \exp\left(-\frac{(x - \mu)^2}{2\sigma^2}\right) \tag{2.2}$$

2.2.1 Image statistics and normalization

Let $I, I_1, I_2 \ldots, I_N \in \mathbb{R}^{n \times n}$ be 2D images with mean μ and variance σ^2. An image is described by the sum of the signal f and an additive noise component m.

$$I = f + m \qquad I \sim \mathcal{N}(\mu, \sigma^2), \tag{2.3}$$

where the noise m is Gaussian distributed with zero-mean ($\mu = 0$) and variance of one ($\sigma^2 = 1$). Images in the matrix representation are often reshaped to vectors $J \in \mathbb{R}^{n^2}$ for processing.

$$I(i, j) = J(k), \qquad \text{where } i, j \in n \text{ and } k \in n^2$$

Image mean   The mean (resp. average, expectation) $E(I)$ of an image I is defined as the sum over all pixels divided by the number of pixels forming the image. It equals the intermediate intensity value of the image.

$$E(I) = \frac{1}{n^2} \sum_{i=1}^{I} \sum_{j=1}^{J} I(i, j) \tag{2.4}$$

Image variance and covariance   The variance measures the deviation of pixel values from the mean value. The variance of an image I is defined as the expectation of the squared distance between the image and the expectation of the image.

$$var(I) = E\left[(I - E\left[I\right])^2\right] \tag{2.5}$$

The standard deviation, $std(I)$ is the square root of the variance. Moreover, the covariance cov between two images reflects the relation of these images to each other. If higher values of the one image correspond to higher values in the other one, the two images show an equal behavior.

$$cov(I_1, I_2) = E\left[(I_1 - E\left[I_1\right]) - (I_2 - E\left[I_2\right])\right] \tag{2.6}$$

Correlation and Cross-correlation   The correlation is a statistical measure of the linear relationship between two random variables. A high correlation ($cor = 1; cor = -1$) leads to the assumption that there is a strong correspondence between two variables. Whenever one variable is changed, the other variable has also an effect. A low correlation, e.g. ($cor = 0$), implies a weak correspondence between the two variables. Here, one variable's modification does not affect the other one. The correlation $cor(I_1, I_2)$ between image I_1 and I_2 is defined by

$$cor(I_1, I_2) = \frac{cov(I_1, I_2)}{std(I_1)std(I_2)} \tag{2.7}$$

Furthermore, the cross-correlation (CC) describes the correlation between two random variables, here images, while displacing one with respect to the other one. It is a function of pixel indices with respect to the pixel shift. The NCC is independent of the image values since it is the CC divided by the standard deviation for each variable.

$$NCC_{I_1, I_2}(j) = \frac{\sum\limits_{i=-n}^{n} (I_1(i) - E[I_1]) \cdot (I_2(i+j) - E[I_2])}{\sqrt{\sum\limits_{i=-n}^{n} |I_1(i) - E[I_1]|^2} \cdot \sqrt{\sum\limits_{i=-n}^{n} |I_2(i+j) - E[I_2]|^2}}, \tag{2.8}$$

where i is the i-th pixel in the image and j describes the current lag ,i.e. the current pixel shift between the images.

2.2.2 Projection and rotation

The projection is the sum over all pixels along one axis. Multiple projection images of the same object do not have to be equivalent. The orientation of an object can be changed by the rotation operator in order to get a different image. The rotation of an object either in 2D or 3D is the movement of the object in relation to a fixed point.

Theorem 2.2.1 *The projection is defined as the line integral over one dimension. Let $V \in \mathbb{R}^{n \times n \times n}$ be a three-dimensional density distribution of an object*

$$P(x, y) = \int V(x, y, z)dz, \tag{2.9}$$

where $\int dz$ is the line integral along z and (x, y, z) are coordinates in the Cartesian coordinate system.

Theorem 2.2.2 *Let (α, β, γ) be a three angles describing a movement of an object around itself with respect to a specific coordinate axis in a sphere. Each rotation component is described by one matrix with respect to the coordinate axis*

$$R_x = \begin{bmatrix} 1 & 0 & 0 \\ 0 & \cos\alpha & -\sin\alpha \\ 0 & \sin\alpha & \cos\alpha \end{bmatrix} R_y = \begin{bmatrix} \cos\beta & 0 & \sin\beta \\ 0 & 1 & 0 \\ -\sin\beta & 0 & \cos\beta \end{bmatrix} R_z = \begin{bmatrix} \cos\gamma & -\sin\gamma & 0 \\ \sin\gamma & \cos\gamma & 0 \\ 0 & 0 & 1 \end{bmatrix} \tag{2.10}$$

A rotation of an object around a specific convention ZYZ would be expressed by the multi-

plication of the three rotation operators.

$$R_{zyz} = R_z(\alpha) \cdot R_y(\beta) \cdot R_z(\gamma) \tag{2.11}$$

The convention ZYZ is used in this book. In cryo-EM the biological specimen is a rigid object. Therefore, the triple of angles is referred to as *Euler angle*.

2.2.3 Fourier Transformation

The Fourier Transform (FT) is the disassembly of a signal in space (resp. time) domain into a weighted sum of sine waves. The Fourier transformed image is in the frequency domain, i.e. Fourier space. A spatial frequency denotes the number of repeated sinusoidal components per unit distance. Imagine a cake being the measured signal in real space. This cake was created by multiple ingredients analogous is the concept of the projection image which is a combination of transformed electron waves in real space(see subsection 2.3.2). It is possible to evaluate all ingredients of the cake by looking up the cake recipe. Similar to the cake recipe the Fourier Transformation provides the recipe to define the weighted sum of sinusoidal waves, e.g. in Figure 2.1, composing the projection image. Hence, it is possible to modify the sine waves and therefore, the projection image, comparably to modifying the ingredients for the cake.

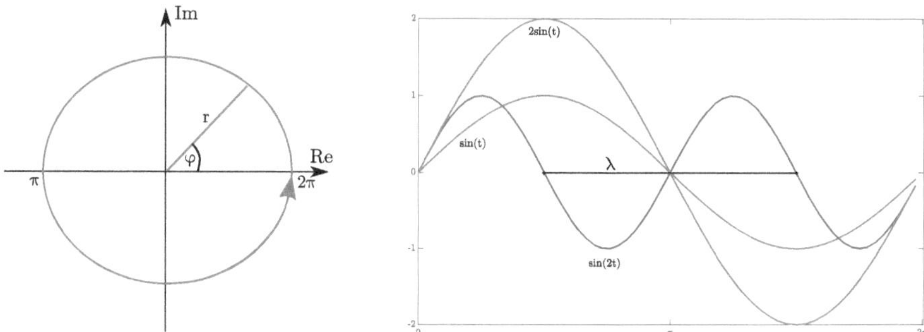

Figure 2.1: Exemplary complex representation Here, the complex coordinate system is shown on the *left*. With the radius r, equivalent to the magnitude, and the angle ϕ, equivalent to the phase, all complex numbers are describable. Three sine waves with different amplitudes and phases are plotted to the *right*. The Fourier image is a combination of multiple sine waves with different amplitudes and phases. A Fourier image breaks down the real space signal into its single components. λ resembles the wavelength of $\sin(2t)$. The sine and cosine waves are periodic.

The advantage is that the signal in space domain can be decomposed into multiple sinusoidal functions with different amplitudes and phases. Low frequencies in the Fourier domain are the representation of large continuous sections of an image (resp. volume). These waves correspond to smaller phases changes. In contrast, high frequencies are related to rapidly changing information with higher phase angles in Fourier space. The zero-frequency component of a Fourier transformed object is called the DC-component. All frequencies are summed up in the DC-component which is equivalent to the complete mass of the object.

Reminder

Euler' Formula (see Figure 2.1)

$$i^2 = -1 \text{ where } i \in \mathbb{C}$$
$$r \cdot \exp(i\varphi) = r \cdot \cos(\varphi) + r \cdot i \cdot \sin(\varphi)$$

Theorem 2.2.3 (2D Fourier Transform) *[[39, Ch.4]]*

$$\tilde{F}_{f(x,y)}(u,v) = \int\int f(x,y) \exp\left[\left(\frac{-2\pi i}{N}\right)(ux+vy)\right] dx dy, \tag{2.12}$$

where $f(x,y), x,y = 0,1,...,n-1$ is an uniformly samples sequence and u,v are the spatial frequencies. i is the imaginary unit.

To compute the Fourier transformation of an object the FFT is often used. It is a fast computational algorithm of the Discrete Fourier Transform (DFT), which is the FT of a discrete signal. A derivation of the FFT is done by Gray & Goodman [39] or Rao *et al.* [60]. The *Fourier Transformation* is reversible without loss of information up to numerical inaccuracy. This is called Inverse Fourier Transform (IFT) and an efficient computation is done by the Inverse Fast Fourier Transform (IFFT).

Theorem 2.2.4 (2D Inverse Fourier Transform) *[[39, Ch.7]]*

$$f(x,y) = \int\int \tilde{F}(u,v) \exp\left[\left(\frac{2\pi i}{N}\right)(ux+vy)\right] du dv, \tag{2.13}$$

where $f(x,y), x,y = 0,1,...,n-1$ is an uniformly samples sequence.

Image processing profits from the properties of the Fourier transformed image. However, important to notice is that the Fourier Transform does not change the properties of the image itself. The Fourier transformed image is only a different representation of the real space image.

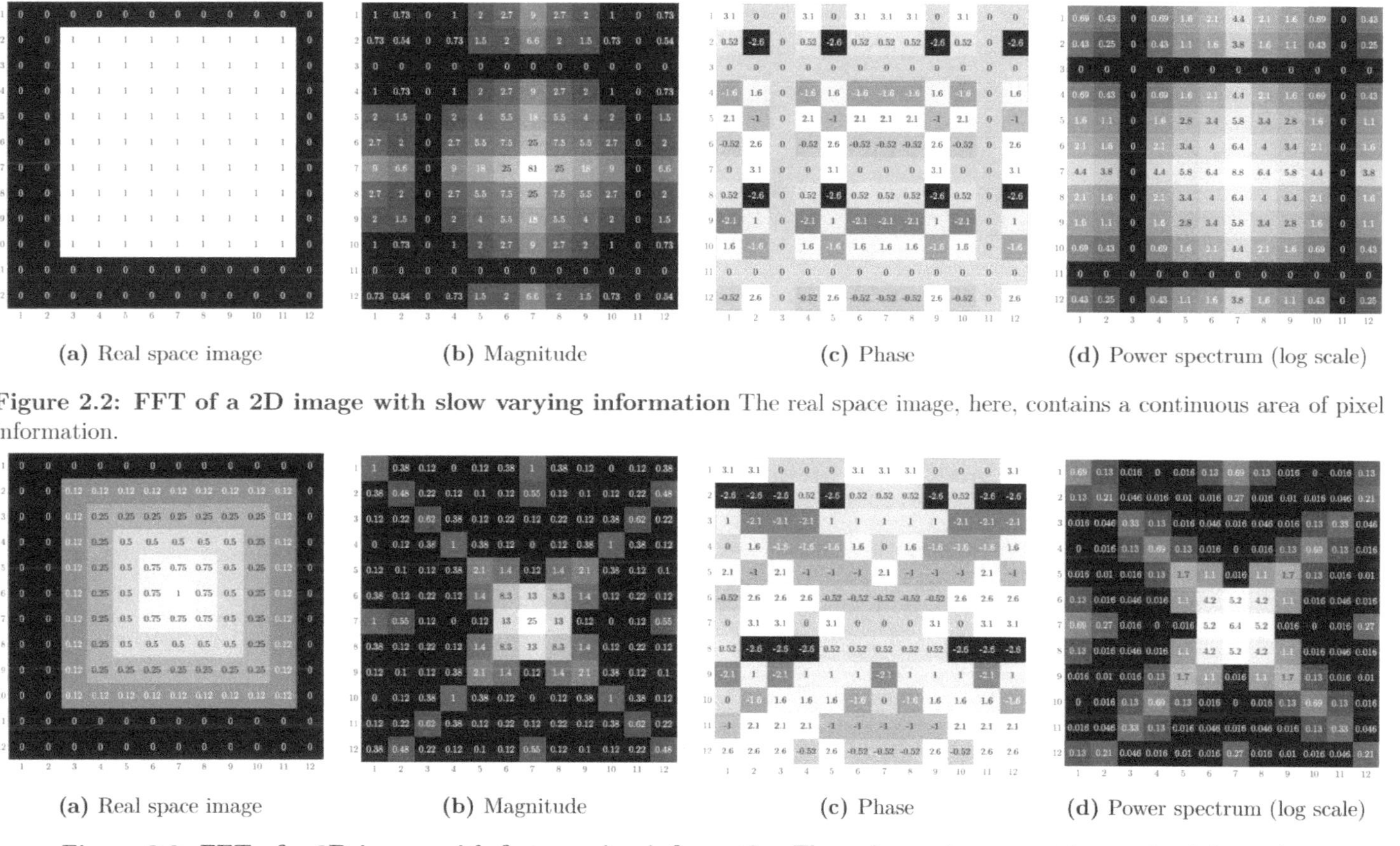

(a) Real space image   (b) Magnitude   (c) Phase   (d) Power spectrum (log scale)

Figure 2.2: FFT of a 2D image with slow varying information The real space image, here, contains a continuous area of pixel information.

(a) Real space image   (b) Magnitude   (c) Phase   (d) Power spectrum (log scale)

Figure 2.3: FFT of a 2D image with fast varying information The real space image contains varying information.

Fourier Properties Each pixel in a Fourier image $\tilde{F}(u,v)$ corresponds to a spatial frequency. This Fourier value is described by a complex number, where $\mathcal{I}(\tilde{F}(u,v))$ is the imaginary component and $\mathcal{R}(\tilde{F}(u,v))$ the real component at spatial frequencies (u,v). The Fourier image is symmetric. [60, Ch.5]

- Magnitude of $\tilde{F}(u,v)$

$$MAG_{\tilde{F}} = |\tilde{F}(u,v)| \tag{2.14}$$

  The magnitude (see Figure 2.2) of Fourier image $\tilde{F}(u,v)$ explains the amounts of spatial frequencies (u,v) present. It has no expressiveness about the direction of the sinusoidal wave.

- Phase of $\tilde{F}(u,v)$

$$\phi = \arctan\left(\frac{\mathcal{I}(\tilde{F}(u,v))}{\mathcal{R}(\tilde{F}(u,v))}\right), \tag{2.15}$$

  where arctan is the arc-tangent. The phase (see Figure 2.2) of a Fourier component denotes the phase shift the sine wave undertook to the non-shifted sinusoidal wave at spatial frequencies (u,v).

- Power spectrum of $\tilde{F}(u,v)$

$$Pow(\tilde{F}_f(u,v)) = |\tilde{F}_f(u,v)|^2 \tag{2.16}$$

  The Power spectrum of a Fourier transformed function is defined as the squared normalized magnitudes. It holds the information about the energy distribution of the frequencies. Especially for noisy data, the power spectrum gives the information, where the ideal signal's sine waves sum up in their intensity. In Figure 2.2 the features of power spectrum are equivalent to the features of the magnitude image.

- Complex conjugate of $\tilde{F}(u,v)$

$$\overline{\tilde{F}(u,v)} = \tilde{F}(-u,-v) \tag{2.17}$$

  The complex conjugate of a FFT is the Fourier value with the same phase but opposite magnitude at spatial frequencies (u,v).

- Linear operator
  The FFT is a linear operation. Let $f,g : \mathbb{R} \to \mathbb{R}$ be two real-valued functions and $a,b \in \mathbb{R}$ then the FFT of the sum of the two functions is equivalent to the sum of the FFT of the two functions.

$$\tilde{F}(a \cdot f(x) + b \cdot g(y)) = a \cdot \tilde{F}(f(x)) + b \cdot \tilde{F}(g(y)) \tag{2.18}$$

- Shift in real space [60]

  A shift in space/time domain results in a phase shift of the FFT. Let $a, b \in \mathbb{R}$ and $\tilde{F}(f(x,y)) = \tilde{F}(u,v)$

$$\tilde{F}(f(x+a, y+b)) = \exp\left[\left(\frac{-j2\pi}{N}\right)(ua + vb)\right]\tilde{F}(u,v) \tag{2.19}$$

- Rotation in real space [61]

  A rotation in space (resp. time) domain results in an equivalent rotation of the FFT. Let $f(x,y) \in \mathbb{R}$ and $\tilde{F}(f(x,y)) = \tilde{F}(u,v)$

$$\mathcal{R}(f(x,y)) = \mathcal{R}(\tilde{F}(f(x,y))) \tag{2.20}$$

Comparing the two real space images in Figure 2.2a and Figure 2.3a the information of the images differentiate in their variation of pixel intensities. In Figure 2.2a there exists one smooth square as compared to the varying informative square in Figure 2.3a. In Fourier space a high variation of the pixels within the real space image is reflected by the higher spatial frequencies (see Figure 2.3). Low spatial frequencies represent smooth regions of the real space image (see Figure 2.2). The DC-component of quadratic Fourier transformed images is on the $(n/2 + 1, n/2 + 1)$-th pixel as seen in Figure 2.2b and Figure 2.3b. Both magnitude and power spectra (see Figure 2.2b, Figure 2.2d and Figure 2.3b, Figure 2.3d) resemble the difference. A great advantage of the Fourier space is that the convolution ($*$) between two functions in real space becomes a multiplication of the two Fourier transformed functions in Fourier space.

Theorem 2.2.5 (2D Fourier Multiplication) *[[60, page 138]]*
Let $f(x,y)$ and $g(x,y)$ be two real periodic sequences with period N along x and y. Let $\tilde{F}(u,v)$ Fourier representation of f and $\tilde{G}(u,v)$ Fourier representation of g be. The convolution between f and g at $(\hat{x}, \hat{y})$ is given by $h(\hat{x}, \hat{y})$ and equivalent to the multiplication of the Fourier transformed functions.

$$h(\hat{x}, \hat{y}) = \frac{1}{N^2} \sum_{x=0}^{N-1} \sum_{y=0}^{N-1} f(x,y)g(\hat{x} - x, \hat{y} - y) \tag{2.21}$$

$$\Leftrightarrow$$

$$\tilde{H}(u,v) = \frac{1}{N^2}\tilde{F}(u,v) \cdot \tilde{G}(u,v) \quad u,v = 0, \ldots, N-1 \tag{2.22}$$

The convolution is denoted by $f(x,y) * g(x,y)$.

A proof of Theorem 2.2.5 has been presented in Rao *et al.* [60]. Another advantage of the Fourier domain is the relationship between the 2D projection images to the corresponding

3D structure map of that object. The theorem defining this relationship is called central-slice theorem. The cs-thm states that there exists an equivalence between the 2D projection image in the frequency domain and a central slice perpendicular to the projection direction in the 3D Fourier transformed volume.

Theorem 2.2.6 (cs-thm) *Let $V(x, y, z) \in \mathbb{R}^{n \times n \times n}$ be a 3D density distribution of an object. Its representation in Fourier Space is given by*

$$\tilde{V}(u, v, w) = \int \int \int V(x, y, z) \exp(2\pi i(ux + vy + wz)) dx dy dz \qquad (2.23)$$

Then a central slice of the Fourier transformed object $\tilde{V}$ is equivalent to the Fourier transformed projection through V. The slice is perpendicular to the direction of the projection.

$$P_V(x, y) = \int \int \tilde{V}(u, v, 0) \exp(2\pi i(ux + vy)) du dv \qquad (2.24)$$

A proof for this theorem is given by Van Heel & Harauz [37]. An extension of the cs-thm to n-dimension is presented by Garces *et al.* [61].

2.3 Imaging in electron cryo-microscopy

Resolving biological samples as small as protein complexes are difficult. The light microscope is not able to visualize neither proteins nor protein complexes [31, Ch.1]. The wavelengths of visible light are not sufficient to resolve objects with diameters of Ångstrom. The smaller wavelengths of electrons made it possible for Ruska [62] to show the first structure of a protein complex imaged with an Transmission Electron Microscope back in 1941. To understand the image formation subsection 2.3.2 of a specimen in a TEM it is necessary to introduce the microscope architecture and its operating principles.

Even though the focus of the **book** lies on imaging protein complexes and their data processing, a short introduction how to prepare a specimen is given.

Sample prep Initially, the protein complexes are purified. There are two sample preparation methods to prepare macromolecules for imaging in a TEM. E.g. in negative stain, the biological sample is applied onto a continuous carbon film after purification [63]. After absorbing the proteins, the sample is stained by a heavy metal salt. It is important to keep the stain layer to a minimum. Advantages are the high-contrast projection images and the fast preparation of negative stain specimen. Due to the fact that negative stain data is often limited in resolution the resulting low resolution structures (roughly around 20 Å) are used as start models for cryo-EM structure determination. Over the past years another sample preparation method was established. Specimens are prepared by plunge freezing [63]. A solution containing the particles is applied onto a grid which has a thin support film such as carbon. The specimen is rapidly frozen in a cryogenic liquid e.g. liquid ethane. Hereby, the water is kept in a vitreous state, which means that the water is not crystallized [64]. Samples studied under cryogenic conditions have the advantage to keep the macromolecules hydrated. A second advantage is that plunge frozen macromolecules are less sensitive to beam damage.

2.3.1 Transmission Electron Microscopy (TEM)

A microscope is a device to resolve objects which are often smaller than the resolving power of our eyes. In order to make protein complexes visible the TEM is used to image these. It composed of an electron source, multiple lenses, apertures and a detector (see Figure 2.4). In the electron gun, an incident beam is generated. A lens in a TEM alters the beam, whereas the apertures control the amount of passing electrons. Finally, the detector measures the radiation of the transmitted beam.

The *electron gun*, the electron source in the TEM, emits electrons to form a coherent beam under the processes called *Thermionic Emission* or *Schottky Emission* [65, Ch.4]. The electron gun contains a cathode. By applying a potential shift between the cathode and the anode electrons are accelerated. The electrons are accelerated by applying voltages in the range of $100\ keV$ - $300\ keV$ to the surrounding cathode tip. At this point all electrons waves have the identical phase information.

Further, in order to prevent an electron scattering by gas molecules, the column of a TEM is maintained under an ultra-high vacuum [42] using multiple vacuum pumps along the column. The quality of the vacuum is important since the less gas molecules are present the more is the unwanted interactions of electrons with gas reduced.

The *condenser lens system* in a TEM contains multiple lenses to form the illumination area. Mostly, a three condenser lens setup (C_1, C_2, C_3) is used for high resolution electron cryo-microscopy. The lenses (see Figure 2.4) are energized by a current introducing a magnetic field. Through altering the strength of the current, the electromagnetic field of the lens is manipulated and hence the refractive power of the lens is changed. Thus, it is possible to change the intensity of the beam or illuminate the area of the specimen, e.g. at the specimen level. [31, Ch.6]

The front focal plane is located above the objective lens. Here, the electrons of the beam converge to a point, called cross-over, and then emerge parallel to the optical axis. The specimen is inserted between the upper and lower objective lens as seen in Figure 2.4. It is on a fixed position in the column of a TEM. Here, electrons interact with the protein complexes in the vitrified ice. They pass through or scatter elastically or in-elastically. Elastic scattering occurs when an electron is scattered without an energy change. Inelastic scattering means that the scattered electrons pass on some energy to the specimen. The

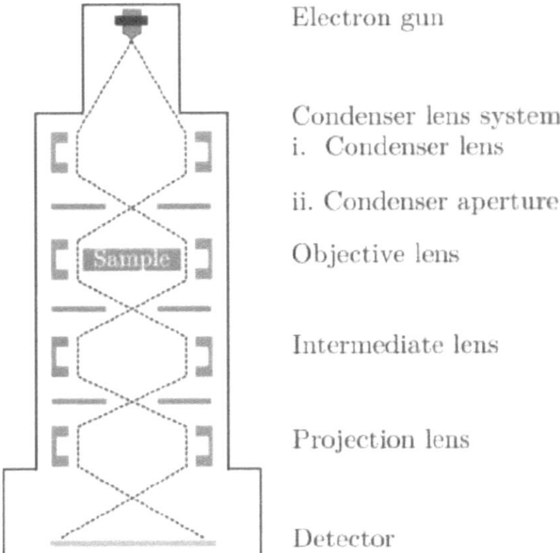

Figure 2.4: Architecture of the Transmission Electron Microscope *From top to bottom*: The electron gun in the TEM is the electron source. Here, the electrons are emitted. The condenser lens system composed of electromagnetic lenses and apertures form the incident electron beam. The projection system of the TEM consist of the intermediate and the projection lens. The detector measures the transmitted electron intensity.

consequences of inelastic scattering can be the ionization of the atoms, X-ray emission or secondary electron scattering as mentioned by Orlova & Saibil [42]. Further details on image formation by electron scattering are presented in subsection 2.3.2.

The objective lens focuses all electron waves in the back focal plane. It produces the optimal projection image of the specimen. By changing the current of this lens the specimen is magnified. The back focal plane is located below the lower objective lens in the diffraction plane (Figure 2.4). Diffraction is mathematically related to the Fourier image.

There are two projection lenses following the objective lens as seen in Figure 2.4. The first one, also called intermediate lens, determines the image being real space projection or a diffraction pattern. The second projection lens is the last lens in a TEM. Here, the image is further magnified.

The aberrations of the lenses in electron optics are similar to light microscopy. There exist the spherical aberration, astigmatism, curvature of a field and coma. Some of these aberrations are able to be corrected with the Contrast Transfer Function detailed in subsection 1.3.1.

The magnified image is detected and digitized. For this purpose, the electron distribution is measured by a florescent screen, which is coupled to a CCD camera as explained by Reimer & Kohl [65]. The CCD camera is not the only way to detect the signal. Nowadays, direct detectors are available. The direct detector in a TEM measures the intensities of the electron waves. The spatial position of the electron is related to the intensity of the electron wave function. The direct detectors are highly sensitive to the energy of an electron since it directly detects the electron.

2.3.1.1 The resolving power of the TEM

A human being, e.g., is capable to differentiate between two points up to a distance of around 0.1 mm. This distance defines the capacity of our eyes to resolve objects and hence, is the resolving power of the human eye [31, Ch.1]. The resolving power of a microscope is the minimal-resolvable distance between two point sources in the imaged object. In general, the resolving power of a microscope depends on the design of the instrument and among other things on its imaging source.

One of the influences here are the wavelengths of the imaging source which range for light from 750 nm up to 380 nm (see Figure 1.2). In theory, the TEM is able to visualize atoms. Due to the fact that electrons used to image in a TEM are much smaller in size as an atom and the relationship between the energy level of an electron and the wavelength discovers by Louis de Broglie's discovered (see Figure 1.2). [31, Ch.1]

$$\lambda = \frac{1.22}{\sqrt{\mathcal{E}}}, \tag{2.25}$$

where $\mathcal{E}$ is the energy of an electron in eV. Here, the wavelength λ is given in nm. The wavelength of an electron depends on the energy level of that electron. In Figure 1.2, the wavelength for a 300 keV electron is stated to be about 1.96 pm. Therefore, the TEM is capable to resolve small distanced points of the object sources. However, the true resolving power of the TEM is influenced by the quality of the lenses and the mechanical stability. [31, Ch.1]

2.3.2 Image formation

One of the main differences cryo-EM to other structural determination methods is that the output of the TEM is a projection image. The micrograph is a true projection of the *Coulomb potential* of the protein complexes [66]. The TEM like any other imaging instrument needs to generate image contrast to make objects distinguishable. Image contrast is the difference in intensities in an image. In a TEM this image contrast is formed by the wave interference of the incident beam and the electrons scattered by the protein complex in the specimen plane. Therefore, the mathematical representation of the single emitted electron and its detected properties as well as the interaction with other electrons are important. The detected interference pattern describe the signal of the protein complex which is read out to a digital 2D representation of the sample. This image is called phase contrast image.

The electron gun of a TEM (see Figure 2.4) emits electrons with a specific magnitude and phase. The electron in the incident electron beam in Figure 2.5 is expressed by the wave Ψ. The wave describes the movements of the electron.

$$\Psi_0 = \psi_0 \cdot \exp(i\phi), \tag{2.26}$$

where the amplitude is expressed by ψ_0 and ϕ is the phase of the wave Ψ. It is an oscillatory function. An equivalent representation is the sinusoidal as seen in 2.1. The wavelength is defined by the smallest distance of two points on the wave with the same phase. Within the column of the TEM the single electron travels through the column, where its path is described by the wave Ψ. In a perfect microscope, the properties of the electrons waves emitted do not change until the sample plane. On the sample plane there are two main possible effects visible. Some electrons pass through the specimen without interaction as the black solid arrow in Figure 2.5. Their wave representation properties Ψ do not change. Other electrons interact with parts of the specimen. These electrons are scattered by atoms. In general, electron scattering underlies different scattering processes. The elastically scattered electrons are electrons which undergo a change in the direction of propagation. This results in a phase shift in the electron wave Equation 2.26. The image contrast in the TEM is based on the interference of the unscattered electrons with these elastically scattered electrons. The exit wave Ψ_{ex} (in Figure 2.5) at a position $\vec{r} = (x, y, z)$ in the specimen plane is described by

$$\Psi_{ex}(\vec{r}) = \underbrace{\psi_0 \cdot \exp\left(i\phi + i\rho\ Pt_{pr}(\vec{r})\right)}_{=\psi_0 \cdot \exp(i\phi) \cdot \exp(i\rho\ Pt_{pr}(\vec{r}))} \tag{2.27}$$

$$\Psi_{ex}(\vec{r}) = \Psi_0 \cdot \exp\left(i\rho\ Pt_{pr}(\vec{r})\right), \tag{2.28}$$

where Ψ_0 is the incident wave emitted from the electron gun in Figure 2.5. The variable $\rho = \frac{m_e \lambda}{h^2/2\pi}$ is defined by the Max Planck constant h, the wavelength λ and the electron mass. Pt_{pr} is the integral of inner potential of the specimen

$$Pt_{pr}(\vec{r}) = \int_{-t/2}^{+t/2} Pt(\vec{r}) dz, \tag{2.29}$$

where it is integrated over the thickness t of the sample along the optical axis z. The phase shift $\Phi(\vec{r})$ the exit wave is equal to $\Phi(\vec{r}) = \rho \int_{-t/2}^{+t/2} Pt(\vec{r}) dz$. The combination of the exit wave of the scattered electrons and the wave of the incident electron beam describe the interference between the electrons. Under the assumption of the weak phase approximation, meaning the changes of the phase are close to zero, the interference pattern is described by the following exit wave [67]

$$\Psi_{ex}(\vec{r}) \approx \Psi_0 \cdot \underbrace{(1 + i\Phi(\vec{r}))}_{\text{with the Taylor series}}, \tag{2.30}$$

41

The first part of this Equation 2.30 is defined by the unscattered electrons (see the black arrows in Figure 2.5). The scattered electron wave (represented by blue arrows in Figure 2.5) are considered by the second additive term in the Equation 2.30.

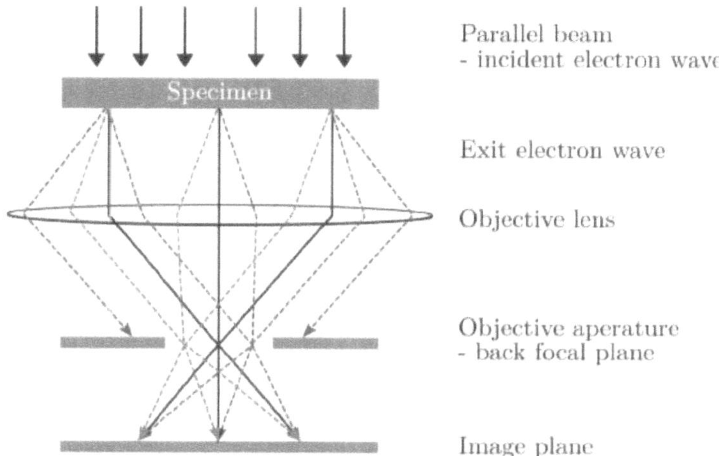

Figure 2.5: Image formation in a TEM An incident electron beam is emitted from the electron gun and passes through the condenser lens system. This wave reaches the specimen sample plane with specific in sync wave properties. The solid black arrows is the direct beam. These electrons were transmitted without an interaction with the particles. The dotted arrows represent the electrons which underwent a shift in their phase due to deflection. The objective lens focuses all waves contained in the beam in the back focal plane. Electrons scatter with too high phase angles as the purple arrows are absorbed by the objective aperture. The back focal plane bundles all electrons with the same wave function properties to a single point. The bundled points equal the diffraction pattern. In the image plane the image intensity is measured.

In Figure 2.5 all electrons with the same scattering angle are bundled on back-focal plane by the objective lens. Each spot on the back-focal plane corresponds to a spatial frequency in diffraction plane. Higher (resp. lower) spatial frequencies correspond to higher (resp. lower) scattering angles. The higher the scattering angle the further distanced from the center is the diffraction spot on the back-focal plane. Too high scattering angles are absorbed by the objective aperture as seen in Figure 2.5. Besides elastic scattering, other scattering processes such as inelastic scattering occur. The electron interacts with the electron shell of the atom in the protein complex and undergoes a phase shift and additionally a change in energy. Inelastic scattered electrons contribute to noise [41].

The position probability density function describes the likelihood to find an electron at a given position. The function is given by the multiplication of the wave function Equation 2.30 with its complex conjugate (see (2.17)). The measurable image intensity [67] based on the position probability density function for Equation 2.30 is

$$I(\vec{r}) = \underbrace{1 - (i\Phi(\vec{r}))^2}_{\Psi_{ex}\cdot\Psi_{ex}^*} \tag{2.31}$$

Keeping the characteristics of the exponential function in mind, thin organic samples such as proteins will be imaged with weak contrast. The phase shift is not measurable. Protein complexes are not visible within the projection image. Therefore, the scattered wave is shifted by an additional phase shift of 90 degrees in order to convert an initially small phase shift into a large change in amplitude [67]. This phase shift is introduced by defocusing the objective lens. The digitized image equals the position probability of the electron. The output of the TEM is called micrograph. It contains hundreds of imaged single particles. During image processing these particles are identified and cut-out for the image processing. The i-th projection image I_i is the integral over the Coulomb potential of the particles. The convolution of PSF and *envelope function* with the effective potential ensures the correction of some aberrations. The mathematical real space representation of the detected signal of a single particle is given as

$$I_i(x) = PSF_n(x) * e_n(x) * \left(\int PT(T_i(\vec{r}))\, dz + m_i^S \right) + m_i^B, \tag{2.32}$$

where T_i describes the unknown rotation angles (α, β, γ) and translation (x, y) of the single particle shown in the i-th image. The term m_i^B is the colored background noise and m_n^S describes the noise due to scattering by the support film. They describe the amount of noise present in the projection image. Both, m_i^B and m_i^S are assumed to be Gaussian distributed with zero mean, statistically uncorrelated and independent.[66]

A simpler real space representation of the cryo-EM projection image is

$$I_i = f + m_i \quad m_i \sim \mathcal{N}(0, \sigma^2) \tag{2.33}$$

and its corresponding Fourier representation is

$$G_i = F + M_i, \tag{2.34}$$

where f is the signal in real space and F is the Fourier transformed signal. m_i and M_i are the corresponding zero-mean, uncorrelated Gaussian noise components in the i-th image in real and Fourier space. These two equivalent representations are the foundation of the image processing tools. The algorithm used to refine the data assumes the image to be a model of the protein signal disturbed by an additive random process (see subsection 1.3.2).

2.4 Single Particle Analysis (SPA)

SPA uses thousands of projection images of the same protein complex to reconstruct the structure of that complex. Each projection image is a representation of that particular protein complex with a specific orientation. The image formation was explained in subsection 2.3.2. To process single particle projection images it is essential to identify, select and cut-out the protein complex on each micrograph. These images are stacked and transferred to the image processing software tools. The aim of signal processing, in general, is to average thousands of similar single particle images to reduce the noise component in the measured signal. The noise reduction leads to an enhancement of the SNR and a more straightforward evaluation of the data.

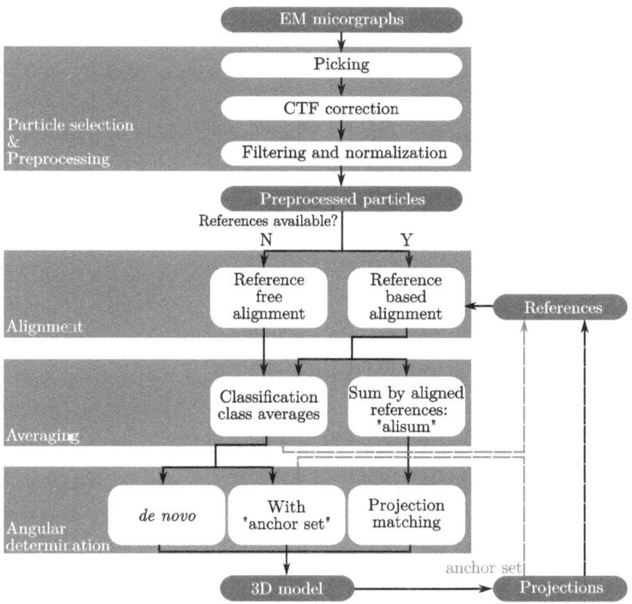

Figure 2.6: Work flow of image processing in single particle analysis SPA is sectioned into four main processing steps. After preprocessing, there are two different ways to align the data. In general, the data is aligned at first and afterwards averaged to build class sums with a better SNR using clustering algorithms. These averaged images are used to determine the unknown projection angles of that image. By applying back-projection algorithms the 3D density maps of the protein complex is constructed. The map is the new reference for the next alignment routine. This figure is taken with the courtesy of Wen-ti Lu and adapted.

At first the particles are identified on micrographs (see Figure 2.6). This data is corrected for the aberrations of TEM followed by a filtering and normalizing step. Processed particles as seen in Figure 2.6 need to be aligned in order to be averaged over the similar projection images. There are two possible algorithmic solutions depending on the availability of references. Following the alignment is the classification. The sum over classified single particle projection images is done. The averaging step ensures to increase the SNR of the projection images. The noise is distributed with zero mean and therefore, reduced by averaging. Consequently, the signal of the class sums of the protein complex is enhanced so that the degrees of freedom (see subsection 1.2.1) for these images can be determined. The classification can also be an intermediate step to produce references for a new alignment.

In Figure 2.6 three different mathematical concepts, i.e. the *de novo* approach, the *anchor set* and *projection matching* are named. All three methods calculate angles for each particle projection image. The aim is then to back-project these images and construct a 3D model of the protein complex. This map is projected with a fixed angular sampling. These projection images are the reference images for the next iteration. The processing restarts with the alignment step in Figure 2.6. Iteratively, the map is refined.

From this point on let $N \in \mathbb{N}$ describe the number of projection images in the data set. Let $n \in \mathbb{N}$ be the number of pixel along one dimension. A pixel is one entry in a 2D image. When the term voxel is used, the 3D equivalent of a pixel is meant. In this book σ^2 denotes the variance and σ the standard deviation of an image.

2.4.1 Preprocessing

Preprocessing in Figure 2.6 is the standardization of single particle projection image properties. The images result from multiple micrographs incorporated with different aberrations and gray values which makes it difficult to process the data as complete set. At first the projection images are CTF corrected (see subsection 1.3.1) meaning to correct some of the aberrations of the TEM. After this the images are filtered and normalized. Image filtering intents to decrease the noise component and remove gradients on the projection images. Normalization is a consequence on the assumption that all projection images belong to the same specimen but have been imaged from multiple grids onto independent micrographs with different gray values. Therefore, it cannot be assured that the level of contrast is uniform over different micrographs. To reconstruct the single particles by averaging they need to be normalized to an even level of gray values.

Image filtering The image is Fourier transformed and then multiplied by a filter function, e.g. Gaussian or cosine shaped functions. All Filter functions can be categorized into Low-pass, High-pass or bandpass. Low-pass filters, e.g. rectangular function, let all low frequencies up to a certain threshold pass. In contrast, the high-pass filter keeps all frequencies of a signal higher than a specific cut-off. The third category is called bandpass filter because it has two cut-off levels to pass frequencies higher than the low cut-off frequency and lower than the high cut-off frequency. In general, filtering has a greater effect on the higher spatial frequency information and removes fine features of the protein complex.

Filtering single particle projection images aims to reduce the noise-related information by reducing higher spatial frequencies. Detector noise (see 1.3.3) being Gaussian distributed is capable to be removed by spatial frequency filtering.

Image normalization Sorzano *et al.* [68] discusses multiple normalization strategies. Here, the most common one is defined. Let I be a projection image with additive noise as described in (2.3). The normalized projection image $\hat{I}$ is

$$\hat{I}(x,y) = \frac{I(x,y) - E(I)}{var(I)} \tag{2.35}$$

In the normalized projection image $\hat{I} \sim \mathcal{N}(0,1)$ the pixel are normal distributed with zero-mean and the variance equal to one. This process is called standardization.

2.4.2 Alignment

Projection images with a good SNR are essential for the projection angle determination. The CTF corrected, filtered and normalized projection images I_n still have a poor SNR. The noise in the images is normal distributed with zero-mean. Therefore, averaging the projection images with the same orientation results in a reduction of noise and hence, improves the SNR. In order to sum up projection images they have to be aligned and classified (see 2.4.3). Hence, their in-plane rotation and translation is determined. Here, three of the five degrees of freedom of the ill-posed reconstruction subsection 1.2.1 problem are computed. There are two algorithmic options to align the data.

Reference-free alignment In case where there is no 3D reference model available, an initial reference image is generated by calculating a rotational average of the data set. All projection images are summed up and then in-plane rotated to build an average of the rotated mean images.

$$I_{rotAvg} = \frac{1}{360} \sum_{i=1}^{360} R_i \cdot E(I), \qquad \text{with} \quad E(I) = \frac{1}{N} \sum_{n=1}^{N} I_n \tag{2.36}$$

where R is the 2D rotation operator which describes the in-plane rotation as defined in Theorem 2.2.2 for 3D. Since all projection images contain the same specimen, the rational average should show a circle indicating the size of the particle. This is used to center the projection images to the size of the protein complex.

Reference-based alignment A 3D reference map of the protein complex is projected by a defined angular sampling. These 2D reference images are used to match the projection images of the protein complex. Hereby, the projections are rotated and shifted to fit their orientation to the orientation of the best matching reference image. The matrix T in (2.37) describes the in-plane rotation and translation for an image.

$$
T(\alpha, x, y) = \begin{pmatrix} \cos\alpha & -\sin\alpha & x \\ \sin\alpha & \cos\alpha & y \\ 0 & 0 & 1 \end{pmatrix},
\tag{2.37}
$$

where α describes the in-plane rotation and x, y are the shifts along the image axis. Since the true rotation and translation for the projection image are unknown the distance between a reference image and the transformed single particle images is minimized.

$$
\min_{I_{Ref}} \|TI_i - I_{Ref}\| \qquad \forall\, T(\alpha, x, y),
\tag{2.38}
$$

where I_i is the i-th projection image and I_{Ref} is a reference image. The optimization can be either measured through the euclidean distance or a cross-correlation between the two images (see subsection 2.2.1). When an optimal set of in-plane rotation and translation (α, x, y) with respect to a reference image I_{ref} is found the parameters are applied to the original filtered and normalized particle images.

The alignment does not change the SNR. It is the preliminary step for either the classification or the sum over all aligned images. In this step the SNR is increased.

2.4.3 Classification

The intention of a classification in cryo-EM is to identify particles with the same features in order to calculate class sums with a better SNR. In general, classification is the partition of the data into subgroups, in which image properties are alike. Features, e.g. the width of a protein, are expressed by a distance between two or multiple images in a subgroup. Additionally, classification algorithms detect images containing broken particles, ice crystals or pure noise. These are sorted into different subsets than particle classes. To analyze all images with one another the classification becomes an exhaustive search for similarities. A necessary step is the data reduction by e.g. *Principal Component Analysis (PCA)*.

The PCA is a statistical approach where principal components of an orthogonal transformed data set are calculated. The first principal component explains the largest variability in the data. The second component expresses the second largest variability in the data and so on. In SPA principal components are calculated by a *singular value decomposition* of the data. A unique linear combination of principal components describes each projection image in the data set. The optimal set of principal components is chosen to compress the cryo-EM data set.

Following this data compression the data is clustered into subsets. The aim of clustering is to minimize the distance within a subset and maximize the distance between the different subsets. There are different clustering algorithms available. The most prominent ones are *k-means* [69] and *hierarchical clustering*. After clustering, all projection images within a class are summed to a class sum. The class sum image has an increased SNR. To ensure good class sums for further processing the user selects classes with broken particles or random noise and remove these from the data set.

The output of a classification is used for a new alignment step. This procedure is done multiple times until the class sums stabilize themselves. These stabilized class sums are used to reconstruct a 3D model.

2.4.4   Angular determination and reconstruction

As mentioned in an ill-posed reconstruction section (see 1.2.1) there are five degrees of freedom for each single particle to orientate itself on the grid. After image acquisition the Euler angles and the translations (see Figure 1.6) are encountered in the projection image of the protein complex. These unknown variables are essential to reconstruct the 3D model. Three of them were already determined by the Alignment algorithms. The two missing parameters, the two Euler angles β and γ, from the reference correspond to the Euler angles missing in the aligned image. If there is no reference available, the determination of angles is more complicated.

Projection matching   Here, the projection images are matched to reference images [42]. Hence, it is essential that an initial 3D model of the protein complex exists. This 3D model was ideally projected for the Reference-based alignment. There was a fix angular sampling such that the two Euler angles β and γ are also known. Since each projection image was sorted to a reference image by a cross-correlation during the alignment, projection matching uses these results.

Angular Reconstitution   If there is no reference model available, there are algorithmic options to define the two missing variables β and γ. By the common lines approach and the central-slice theorem it is possible to define the rotational relationship of the projection images to each other with respect to the 3D Fourier volume. In detail, the cs-thm stated that Fourier transformed 2D projection images correspond to a slice in the Fourier volume (see Figure 2.7). If the projection angles of an image are known, it is possible to insert the Fourier transformed projection image into the 3D Fourier object as at the position of the corresponding slice. Assuming that at least two different central slices of two distinct

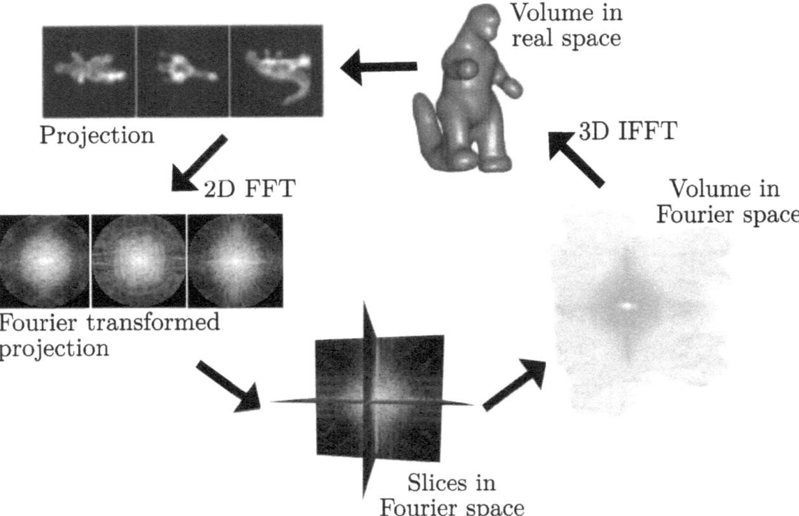

Figure 2.7: Sketch of the cs-thm The cs-thm defines the relationship between 2D Fourier transformed projection images and slices of the 3D Fourier transformed model. The simulated model, a dinosaur, is projected in real space as defined in subsection 2.2.2 and then Fourier transformed (see Theorem 2.2.3). Each Fourier space projection image has a perpendicular slice in the Fourier transformed 3D object.

projection images are inserted in the Fourier volume as seen in Figure 2.7, then there is an intersection between these two planes called common line. If the projection angles are unknown, the projection images still share the common line between their two central slices. The common-line is often determined by computing the NCC. This fact can be used to determine two out of three Euler angles with respect to the rotation $R_{I_2}^{-1} \cdot R_{I_i}$ [29]. This is called common-line approach.

The third missing angle is the angle between these two central slices. In order to determine this a third Fourier transformed projection image is necessary. In Figure 2.7 three projections are shown. Each pair has a common line. All three common lines define the set of Euler angles [29]. By Fourier transforming all projection images in the data set and then computing the cross-correlations the common-lines are determined computed. After inserting all projection images, the 3D Fourier volume is IFFT. It is important to keep in mind that the projections, which correspond to a reflection of the complex, are not uniquely determined. Indeed, it is not possible to differentiate between them.

All these steps (Preprocessing, Alignment, Classification and Angular determination and reconstruction) aim to reconstruct the structure of the protein complex. Often the first round of image processing gives only a low resolution representation of the macromolecule. The aim of electron cryo-microscopy is, however, the reconstruction of high resolution structures. Hence, it is necessary to do multiple rounds of these image processing steps (2.4.1, 2.4.2, 2.4.3, 2.4.4), where the current estimate is used as a reference for the next iteration in Figure 2.6. There are multiple software packages such as *CowSuite* available to refine single particle cryo-EM data.

2.4.5   RELION refinement

The software package *RELION 3.0* by Zivanov *et al.* [24] approaches the determination of the five degrees of freedom of single particles using a maximum likelihood approach. It is based on the mathematical concept of Bayesian statistics which leads to iteratively maximizing the posterior distribution of the five degrees of freedom [70].

$$\hat{\Theta}_{MAP} \in \underset{\Theta}{\operatorname{argmax}} \; \underbrace{P(X|\Theta,Y)}_{Likelihood} \times \underbrace{P(\Theta|Y)}_{Prior}, \tag{2.39}$$

where $\hat{\Theta}_{MAP}$ is the set of missing parameters, namely the Euler angles (α, β, γ) and the translation x and y. Each iteration maximizes Θ to find the best fit of the multiplication between the likelihood of the projection images X and the prior distribution given the current parameter set. Word-for-word it is how likely the projection images represent these parameters (2.39). Additionally, the Bayes approach defines a prior distribution defining a certain requirement to the data.

The following equations are the underlying mathematical concept of RELION published by Scheres [70]. In each iteration the volume (2.40), the noise model (2.41) and the prior (2.42) are maximized.

$$V_l^{(n+1)} = \frac{\sum\limits_{n=1}^{N} \int_\phi \Gamma_{i\phi}^{(n)} \sum\limits_{j=1}^{J} P_{lj}^{\phi^T} \dfrac{CTF_{ij} I_{ij}}{\sigma_{ij}^{2(n)}} d\phi}{\sum\limits_{n=1}^{N} \int_\phi \Gamma_{i\phi}^{(n)} \sum\limits_{j=1}^{J} P_{lj}^{\phi^T} \dfrac{CTF_{ij}^2}{\sigma_{ij}^{2(n)}} d\phi + \dfrac{1}{\tau^{2(n)}}}, \tag{2.40}$$

where i, j are the Fourier components in 2D resp. 3D. The next volume $V^{(n+1)}$ is the sum over all images back-projected after a CTF correction. After each update of the volume, the noise model is optimized.

$$\sigma_{ij}^{2(n+1)} = \frac{1}{2} \sum\limits_{i=1}^{N} \int_\phi |I_{ij} - CTF_{ij} \sum\limits_{l=1}^{L} P_{jl}^{\phi} V_l^{(n)}| d\phi \tag{2.41}$$

The prior information also needs to be updated.

$$\tau_l^{2(n+1)} = \frac{1}{2}|V_l^{(n+1)}|^2 \tag{2.42}$$

Each image $\Gamma_{i\phi}^{(n)}$ is a quality factor for the angular determination. Γ in n-th iteration, i-th image, angle projection ϕ soft assignment of Euler angles [29].

$$\Gamma_{i\phi}^{(n)} = \frac{P(I_i|\phi, \Theta^{(n)}, Y) \cdot P(\phi|\Theta^{(n)}, Y)}{\int_{\phi'} P(I_i|\phi', \Theta^{(n)}, Y) \cdot P(\phi'|\Theta^{(n)}, Y)d\phi'} \tag{2.43}$$

Subsequently to each iteration, the resolution of the current map estimate is calculated. As long as the resolution improves, the next iteration is started. The resolution is defined by the FSC of the two half-maps of the data. If the resolution is constant over multiple iterations, the angular step size is decreased, i.e. more reference projections are considered. In order to terminate the refinement RELION considers the FSC which explains how well the two structural half maps of the gold-standard refinement correlate with each other within each shell. If the resolution number does not change for a couple of iterations, the data has converged to a 3D map of the protein complex.

2.5 Map assessment

Another important concept of resolution is the meaning of feature resolution. To emphasize again, the The resolving power of the TEM is not equivalent to the feature resolution. In general, the feature resolution means up to which point there are distinguishable features present in the map. Common assessment algorithms are the Fourier Shell Correlation (see subsection 1.4.2), the Differential Phase Residual [71] and the Spectral Signal-to-Noise-Ratio [72]. All three of these methods measure the map based on the sine waves present in Fourier space. They compute the spatial frequency up to which features exist in the map. The resolution of a 3D reconstructed cryo-EM map is defined by the smallest spatial frequencies where there are features present. In other words, it is the spatial frequency that corresponds to the fastest traveling sine wave which in turn has the shortest distance between the peaks. The resolution assessment of a cryo-EM map is a critical step in SPA.

2.5.1 Spectral Signal-to-Noise Ration (SSNR)

The SSNR was first introduced by Unser *et al.* [72] to define the resolution based on the particles in cryo-EM. It evaluates consistency of the entire picked particle data set [38]. In general, it is defined as the power of the signal τ_S^2 divided by the power of noise τ_N^2 of the projection image.

$$S = \frac{\tau_S^2}{\tau_N^2}$$

An SNR equal to one corresponds to an equal amount of signal and noise in the image. A value below one equals to a greater amount of noise than signal in the data. The SSNR is also a measure of spatial resolution and therefore, is a function of spatial frequencies.

Theorem 2.5.1 (Spectral Signal-to-Noise-Ratio [38]) *Let there be N images in the cryo-EM data set with the Fourier transformed image I_k. The rotational averaged spectral variance ratio S is given by*

$$S(s) = \frac{\sum\limits_{\substack{||s_k|-s|\leqslant\epsilon}}^{k_s} \left| \sum\limits_{k=1}^{N} I_k(s) \right|^2}{\frac{N}{N-1} \sum\limits_{\substack{||s_k|-s|\leqslant\epsilon}}^{k_s} \sum\limits_{k=1}^{N} \left| I_k(s) - E(\tilde{I})(s) \right|^2}, \tag{2.44}$$

where $E(\tilde{I})$ is the mean value over all images. Here, $S(s)$ is an equivalent estimate of the estimated signal and the noise residual. It is biased with respect to these two components. Therefore, the SNR becomes the SSNR.

$$SSNR(s) = \begin{cases} S(s) - 1, & if\ S(s) > 1 \\ 0, & if\ S(s) \leqslant 1 \end{cases} \tag{2.45}$$

then the SSNR is an unbiased estimate as derived by Unser et al. *[72].*

The SSNR of the sum of the images is equivalent to the SSNR of a single image multiplied by the number of images within the set.

$$SSNR_{E[\tilde{I}]} = \frac{|NF|^2}{E\left[\sum\limits_{n=1}^{N} M_n\right]} = \frac{N^2|F|^2}{N\sigma^2} = N \cdot SSNR_{I_k} \tag{2.46}$$

Similar to the FSC the SSNR has a specific cut-off value up to which spatial resolution of the data is reliable. A conservative threshold of the SSNR derived to be level, where the SSNR drops below the value four within a shell s $(SSNR(s) \sim 4)$ [72]. Here, the resolution is specified by the empirical cut-off frequency $f_4 = \frac{1}{d_4}$ and which in turn provides a measure of the resolution (d_4) of the data set. This threshold was based on confidence interval of the theoretical SSNR [72]. Unser *et al.* [72] derived the distribution of the SNR. A second threshold was set to zero, where in Fourier space no signal is present. Unser *et al.* [72] qualifies the minimal acceptable threshold based on the non-central F-distribution. The review paper by Sorzano *et al.* [73] also introduced the derivation of the non-central F-distribution for the SSNR.

In the end SSNR is biased by the noise component present in the data. The noise model established in the cryo-EM data is based on the normal distribution. The SSNR relies on the claim that the noise is zero-mean. In reality, the noise is not removed by the sum over all images. Therefore, there is still noise present in the average signal [72].

2.5.2　The connection between SSNR and FSC

The relationship between the SSNR and FSC has been discussed in various papers (see Penczek [38], Unser *et al.* [72], and Sorzano *et al.* [73]). Bershad & Rockmore [74] derived an unbiased estimator for the SNR between two Gaussian signals from the normalized CC under the assumption of zero mean and independent noise components. The power spectrum of the noise over the observed data is of equal power [74]. The connection between the FSC and the SSNR was based on statistical behavior of the two computational methods.

$$FSC = \frac{SSNR}{SSNR + 1} \qquad SSNR = \frac{FSC}{1 - FSC} \qquad (2.47)$$

In order to ensure an unbiased model refinement the data is split into two halfes. By this gold-standard refinement the link of the FSC and SSNR becomes the following (see Unser *et al.* [72]).

$$FSC = \frac{SSNR}{SSNR + 2} \qquad SSNR = \frac{1}{2} \cdot \frac{FSC}{1 - FSC} \qquad (2.48)$$

Threshold discussed in subsection 1.4.2 can be related to the SSNR. The SSNR is equal to two, when the FSC curve drops below 0.5. At this point twice as much signal as noise is present in the data. In the field of cryo-EM it has been established that the reolsution of gold-standard processed data is measured at the point, where the FSC drops below 0.143 (see subsection 1.4.2). At this point the SSNR of the data has around three times the power of noise compared to the single power present based on the relation between the FSC and SSNR. The consequence is that the projection images have a high level of noise.

Chapter 3

Results

3.1 From nothing to high-resolution

The noise disturbs the detection of the ideal signal of the protein complex. Furthermore, this noise influences the image processing tools. Especially, the variation of the noise from one pixel to the neighboring pixels (see subsection 1.3.2) makes it challenging to separate the higher spatial frequency-related signals from the noise. Additionally, due to the low electron does used to image the protein complex the projection images have a low SNR. The noise power dominates the power of the signal and the algorithmic tools fail to identify the underlying signal. These effects can lead to the reconstruction of noise-related information. Consequently, the evaluation of the cryo-EM data and its resolution is crucial. To estimate the resolution of the protein structure the FSC as presented in section 2.5 is computed between two reconstructions of the protein complex. To empathize again, the FSC is a measurement of the consistency of the data. It is subject to various misinterpretation issues such as the lack of the differentiation between the signal and the noise. If image processing tools are misused or data interpretation is wrong, then the result often is the false interpretation of the resolution of the reconstructed protein maps. The three experiments underline the difficulties of the image processing tools. Furthermore, the FSC in all experiments results in an overestimation of the resolution for the reconstructed maps.

3.1.1 Systematic error within the CTF correction

The CTF corrects single particle projection images for the aberrations of the TEM (see subsection 1.3.1). The CTF is not a unique defined function for the cryo-EM data set. Moreover, due to image acquisition bing done several times with various grids requiring different configuration the CTF is a set of functions depending on the image acquisition settings. Besides a fixed parameter setting, the user influences the quality of the CTF. The defocus, e.g., is a parameter, which is set by the user specific for each micrograph.

The defocus values are noted in the meta-data file and used to fit a CTF model for each single particle image. During the refinement of single particle images the on-the-fly CTF correction is done. The CTF parameters do not change during this image processing step. If the user incorrectly assigns the defocus values to the micrograph, the CTF is miscalculated. The consequence is the inaccurately correction of the aberrations of the TEM. The question is what impact does a falsely CTF corrected projection image has on the refined structure and its resolution.

Design of experiment A cryo-EM data set of the *Thermoplasma acidophilum 20S proteasome* (see section 1.1) was recorded with a TEM called Titan Krios equipped with a C_s-corrector by Prof. Holger Stark. The obtained micrographs were processed based on the concept of SPA described in section 2.4. The single particles projection images were picked with *Gautomatch* and cut-out with a box size 360 and a pixel size of 0.713 $\text{Å}/pix$. Following the picking the CTF was fitted with *Gctf*. The metadata file contained the single particle positions on the micrograph, defocus and spherical aberration. The set of 990, 010 single particle projections were refined in two independent data sets containing half the data following the gold-standard procedure. To start the RELION refinement a reference structure of the protein complex (see Figure 3.1) was filtered to a map with a low resolution of 40 Å (see Figure 3.1). This map is the initial reference 3D Fourier transformed volume to define Euler angles to the cryo-EM data based on the cs-thm. After the first iteration the reconstructed structure Figure 3.1, an intermediate map, becomes the reference for the next iteration. This refinement converged to two half maps of the protein complex based the unchanged estimated FSC resolution.

Defocus correction (see Equation 1.2)

$$\delta f_{ast}(\theta) = \frac{\delta f_u + \delta f_v}{2} + \frac{\delta f_u - \delta f_v}{2} sin(2(\theta - \theta_{ast})) \tag{3.1}$$

RELION does an on-the-fly CTF correction, which means that after each iteration cycle the cryo-EM data is multiplied by the FSC (see Figure 3.1). Consequently, a CTF miscorrection of the identical T20S proteasome recorded data can be initiated by modifying the metadata file. The parameter δf_{ast} corrects for the defocus setting and the influences of the astigmatism of the TEM. Compared to the first refinement the defocus values $(\delta f_u, \delta f_v)$ and the astigmatism angle θ_{ast} (see Equation 1.2) were shifted. The defoci δf_u, δf_v are moved as a pair along the same direction. The offset for this is shown in Figure 3.2. There is a shift of defocus values for 85 % of the single particle projection images. The other 15 % were displaced to the identical values. On the contrary, the angle θ_{ast} was moved independently of the other two variables. The differences between the original defocus values and the shifted defocus values depending on the sorted defocus δf_u along one axis are shown in

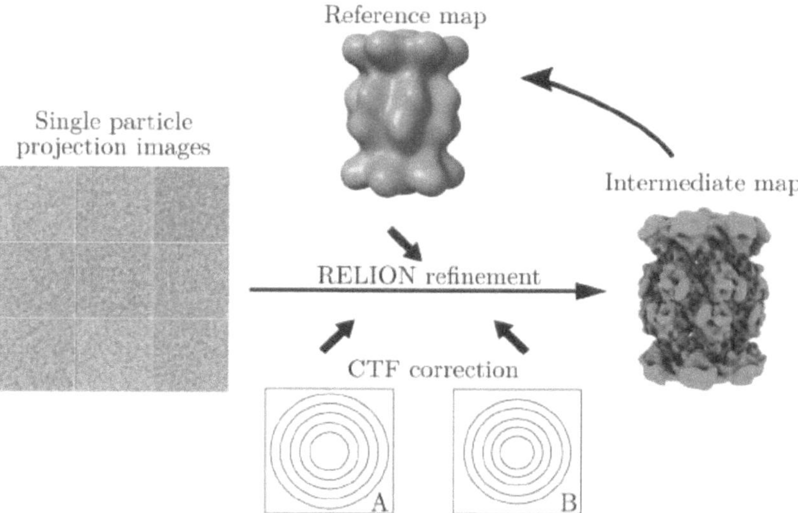

Figure 3.1: RELION workflow of the experiment The single particle projection images are on-the-fly CTF corrected and reconstructed to a 3D representation of the protein complex. The CTF is modified for the second refinement of the cryo-EM data.

Figure A.2, A.3 and A.4. Important to notice is that the RELION refinement initialized with the identical low-resolution reference in Figure 3.1. Despite the CTF miscorrection of the single particle images the refinement also converged. To emphasize again, the convergence of the refinement is based on the unvarying correlation between the two half maps calculated by the FSC, which is interpreted as resolution.

Observation   RELION defines the stopping criterion of the maximum likelihood computation by the determined resolution of the two maps based on the FSC (see subsection 2.4.5). If there is no gain in the structural point resolution for several iterations, RELION stops the refinement. The conservative threshold 0.5 in Figure 3.3 proposes a resolution of 3.29 Å for the original reconstructed map in 3.4. In comparison, the refined map with the wrong CTF defocus parameter in Figure 3.4 has a resolution of 3.55 Å at 0.5. Furthermore, the common threshold of 0.143 for gold-standard refined data estimates an even higher resolution for both maps, i.e. the original map with 2.88 Å and the altered map around 3.17 Å. According to the FSC curves in Figure 3.3 both refined structures reach the resolution, which is normally high enough to start atomic model building.

In Figure 1.11, specific structural geometrical features high-resolution structures were presented. The point resolution of around 4 Å in Figure 1.11 shows maps that have visible α-helices and β-sheets. A resolution around 3 Å starts to define rings of atoms as seen in Figure 1.11. Thus, the interpretation of FSC for the refined T20S proteasome maps leads

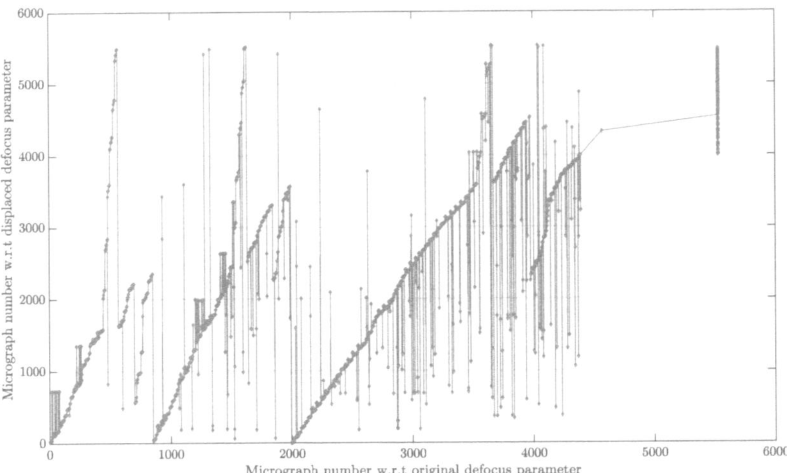

Figure 3.2: Mapping of the displaced defocus values Here, the shift of the defocus values δf_u, δf_v along the minimal and maximal axis (see subsection 1.3.1) are graphed. The value pair $(\delta f_u, \delta f_v)$ was mapped from the original single particle on a micrograph to the new parameter set of another single particle of a different micrograph. One point in the graph corresponds to the mapping from the original micrograph to the new micrograph.

to the conclusion that both structures resolved well and showed detailed chemical structure features. The question is whether the effects of the CTF correction on the reconstruction of the protein complex is less than expected or the FSC does not detect quality issues such as too coarse resolved side chains or missing structures such as α-helices within the map.

A visual analysis of the two refinement maps in Figure 3.4 is done to verify the FSC claimed resolution. Both refinement routines converged to structures with a similarity in their overall geometrical representation. The maps show a cylindrical representation of the protein complex. By previous published T20S proteasome structures in the RCSB PDB this reconstruction, here, shows a similar structure [6]. The map in Figure 3.4b resulted from ideal CTF correction refinement. This second computation depending on the CTF miscorrection converged to the structure in Figure 3.4a. The map in Figure 3.4a does not coincide with the 3D map in Figure 3.4b. Enlarging the identical regions in each map (see Figure 3.4) a visible difference in their fine features becomes apparent. It is easy to identify by eye that reconstructed features in the enlarged sections do not coincide. This contradicts to the assessment by the FSC in Figure 3.3 which resulted in a nominal high resolution for each map.

The protein complex structure in Figure 3.4a does not show the same geometrical representation as the refined structure in Figure 3.4b. Consequently, the maps cannot represent the same conformation of the protein complex. Taking the theory about features of protein

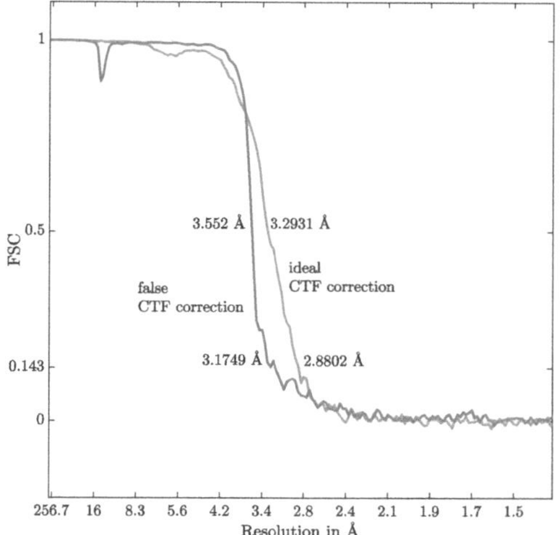

Figure 3.3: FSC curves of the two T20S proteasome refinements Here, the FSC of two refined maps in Figure 3.4 are plotted. The dotted lines represent the two thresholds. Both FSC curves are computed with RELION. The teal graph corresponds to the gray refined map in 3.4. It indicates a resolution of 2.8Å for the map. The pink graph corresponds to the teal map in 3.4, where the wrong CTF correction was done. This FSC curves defines a similarity up to 3.2Å for the map. Both maps reach high-resolution.

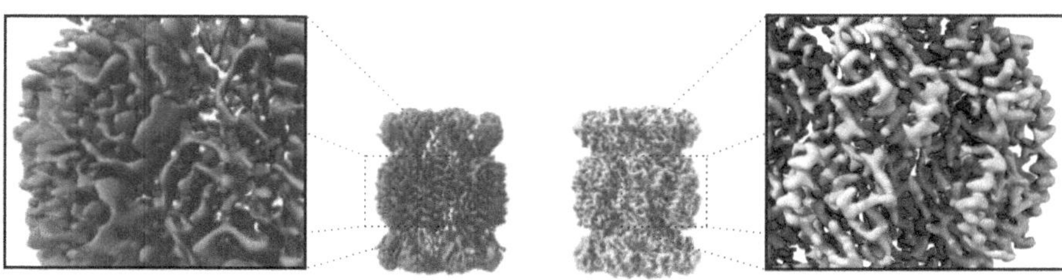

(a) The resolution of the map is around 3.17 Å   **(b)** The map has a resolution around 2.8 Å.

Figure 3.4: Refined structure of the *Thermoplasma acidophilum 20S proteasome* Here, the two maps of the *Thermoplasma acidophilum 20S proteasome* resulting from the two refinements are shown. The microscopic data used to refine both structures was identical. Visually, the structures differ from each other in their fine features. The resolution of each structure is determined by the FSC (see Figure 3.3). Both structures are refined with RELION 3.0 [24].

complex structure at 3Å (see Figure 1.11) into account the enhanced region of the map in Figure 3.4a shows that the features of structure deviate from theory. The protein map in Figure 3.4a does not represent the structural features of protein complexes at 3.17Å. The refined structure in Figure 3.4a is no theoretically accurate protein complex structure. Indeed, the data was corrected with the wrong defocus values during CTF correction. Thus, the wrong phase information were used to reconstruct the structure resulting in a wrong protein complex map. Clearly, the FSC did not detect the quality problems of the 3D density map in 3.4a. Even though in subsection 1.2.1 it was explained that there does not exist a global optimal structure. With the visual deviation from the theory (see Figure 3.4) the protein complex structure is concluded to be a miscalculation despite of the nominal high value of the FSC.

To stress the inaccuracy of the structure the FSC between the two refined T20S proteasome structures in Figure 3.4 is computed. The FSC in Figure 3.5 decreases quickly in the lower spatial frequencies. Lower spatial frequencies correspond to slower varying information such as the overall structure of the protein complex. If the common cryo-EM thresholds are applied, the correlation between these two maps estimates a similarity up to 16.2 Å (reps. 10.39 Å). As described in section 1.4 resolutions, which are lower than 10Å, only indicate a rough estimate of the protein complex structure. Certain features of the protein complex structures are not present at this point. As mentioned above, both T20S proteasome structures show a similarity in their overall representation of the protein complex. The FSC, here in Figure 3.5, supports the overall appearance of the two refined maps in Figure 3.4. This underlines that the estimated high resolution in Figure 3.3 for the T20S proteasome in Figure 3.4a cannot be valid. However, the FSC in Figure 3.5 increases again. This often indicates that the interpretation of the FSC is difficult.

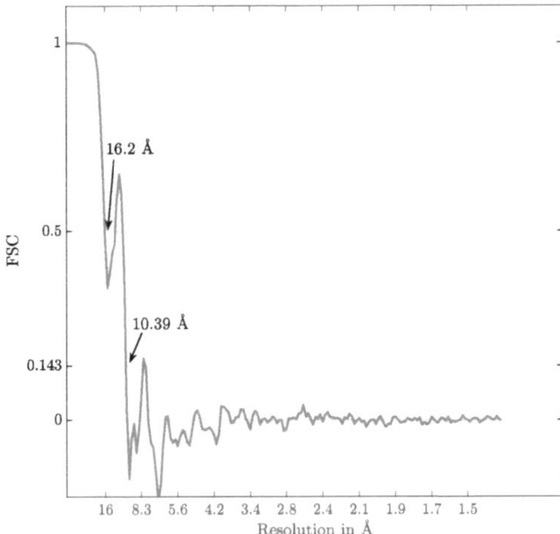

Figure 3.5: FSC between the two differently refined T20S proteasome Here, the FSC between the two refined protein complex maps (see Figure 3.4) is plotted. The FSC drops fast below the common used thresholds. The both structures are assumed to coincide up to resolution of 16.2 Å (resp. 10.39 Å). However, both times the FSC increases again. The reliability of the estimated resolution is doubtful.

This experiment demonstrates the controversy of the interpretation of the FSC. Based on the correlation curves shown in Figure 3.3 both structures contain high resolution features of the protein complex, which contradicts to the visual assessment of the maps in Figure 3.4. There exists a discrepancy between these two evaluations of the reconstructions. Since both structures are reconstructed from the identical microscopic data, the problem is related to the image processing tools.

3.1.2 Fitting noise

Often the poor Signal-to-Noise-Ratio of the projection images leads to poor alignment results. The algorithms as described in section 2.4.2 and section 2.4.2 do not differentiate between the signal and the noise component. Moreover, the reference-based alignment algorithms are based on how well the reference image coincides with the single particle image. To measure the similarity of the images the cross-correlation is computed. Shatsky *et al.* [75] were capable to generate the picture of Albert Einstein from pure noise information through $1,000$ reference-based aligned and summed Gaussian distributed noise images. The next experiment shows how the reference model affects at first the identification of the particles on micrographs as well as the alignment of the data. Moreover, it is shown that these noisy images reconstruct to a replica of the reference protein complex structure.

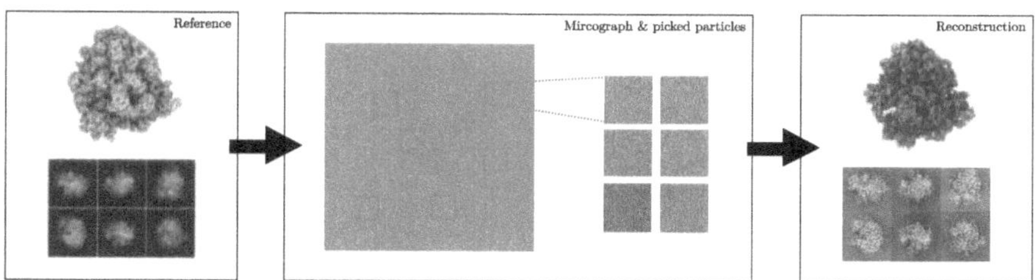

Figure 3.6: From the reference map to the structure *Left*: The reference map of the protein complex called ribosome was used to identify the fake particles on recorded noise micrographs. This map was projected by a certain angular distance. The reference projections were used to identify particles on micrographs such as shown in the middle. *Middle*: The green marked sections represent exemplary identified particle sections. These regions are cut-out and refined. The alignment and reconstruction was done in the *CowSuite*. *Right*: The reconstructed map of the ribosome and its re-projections are shown.

Design of experiment Grids with a carbon-film but no particles and vitrified ice layer were imaged with the TEM by Dr. Jan Erik Schliep[1]. Thus, the recorded micrographs contain pure noise-related information. A density map of the ribosome (see Figure 3.7) with a box size of 420 was generated based on the atomic model (*5afi* in RCSB PDB). The 2Å ribosome was then projected by a specific angular sampling of 3.66 degrees with the *CowSuite*. Six exemplary reference projections are shown in Figure 3.6. A total of 3072 noise-free references of the ribosome are used as a template to identify particles on micrographs. The algorithm implemented in the software package *Gautomatch* identified the particles on the pure noise micrographs as seen in Figure 3.6. With the knowledge of the imaged grids these images do not contain a signal related to the ribosome.

---

[1]Former member of the Department of Structural Dynamics, Max Planck Institute for Biophysical Chemistry

After picking the micrograph regions, the detected non-signal sections were cut-out with a box size of 420 (see Figure 3.6) and split into two random subsets of equal size. Each subset contained 39518 of these normalized non-particle projection images. Both halves aligned to the reference images so that all images were in-plane rotated and shifted with respect to the reference image. Hence, the noise images were matched to a reference projection. The reference images resulted from a forward projection of the 3D model so that the Euler angle pair (β, γ) for each picked projection image is also known. These aligned images were reconstructed by Fourier reconstruction with *CowEyes*.

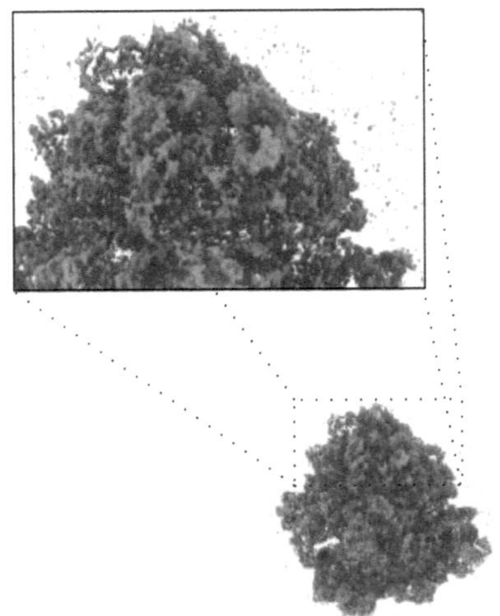

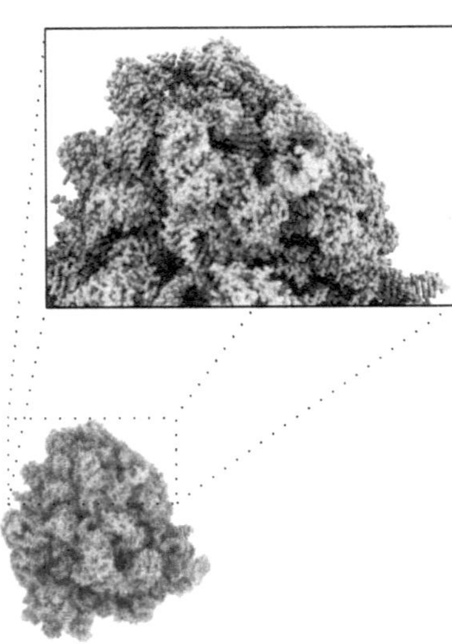

(**a**) Reconstructed map of the non single particles

(**b**) Map of the reference

Figure 3.7: Reference and reconstructed protein complex structures This reference was projected by an angular distance of 3.66 degrees. The map was generated using the published atomic model (RCSB PDB: *5afi*).

Observation   In Figure 3.7 the reconstructed and the reference map of the ribosome are shown. The overall representation of the two maps coincide. The enlarged region of the two maps in Figure 3.7 does also not indicate an obvious deviation between the structure. Within the box of the reconstructed map noise is visible. To emphasize again, the reconstructed map in Figure 3.7 results from pure noise images taken with a TEM. The images did not contain any information related to the protein complex, which nevertheless, is resulting from a Fourier back-projection of 39518 images. This indicated that the reconstructed map should be visually evaluated and interpreted based on details as presented in

Figure 1.11 by an experienced user.

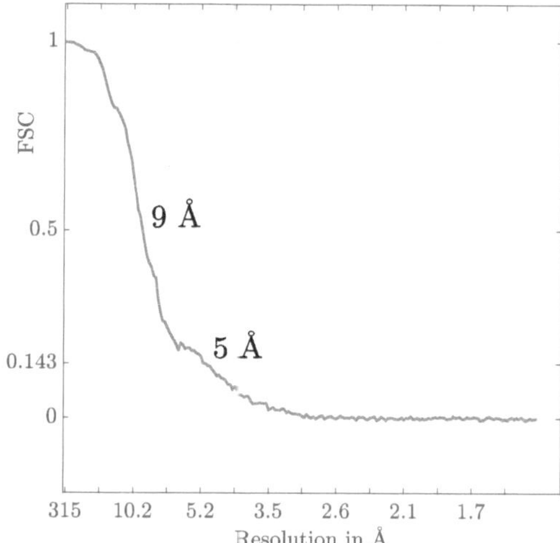

Figure 3.8: FSC of model-biased picked and aligned data Here, the FSC between the two half-set reconstructions (see Figure 3.7) is plotted. The two maps correlate well up to a resolution of about 5Å for gold-standard refined data in Fourier space. In this experiment, the recorded cryo-EM data was not independently picked so the more conservative threshold should be applied. Keeping this in mind the FSC curve estimates the resolution of around 9 Å.

In Figure 3.8 the FSC curve between the two reconstructed half-maps is shown. At the cut-off level of 0.5, the structure is supposed to have a resolution of about 9 Å. Within the shells of these lower spatial frequencies the reconstructed half maps correlate well. With increasing number of the shell the FSC decreases further and drops below 0.143. The estimated resolution of the ribosome at that point is 5 Å. The general descending curve characteristic of the FSC does not indicate issues related to overfitting of noise. The FSC fails again to detect the true quality problems of the map. The particles in Figure 3.9 stress even more the issue between the picked images, the corresponding aligned images and their re-projections of the reconstruction. The recorded data shows no indicator that there exists a signal of a protein complex. The picked and aligned images differ visibly from the re-projection. The re-projection is supposed to be equivalent to the corresponding aligned particle image since it should contain the recorded signal. Within the re-projection there should be always less or equal amount of signal as in the detected projection image.

3.1.3 Adding fake details to a structure

Protein complexes are not rigid objects. They have the capability to move from one conformation to another conformation. These movements cause small changes in their structural representation (see section 1.1). Therefore, a refinement often fails to resolve highly dynamic part at high resolution. In order to push the resolution within these dynamic parts, the classification with respect to these areas is done. The cryo-EM data is classified with respect to reference maps. As the previous experiment has shown that the alignment is

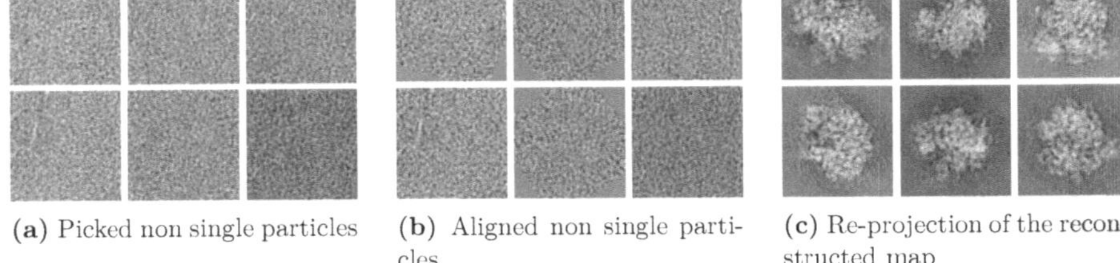

(a) Picked non single particles

(b) Aligned non single particles

(c) Re-projection of the reconstructed map

Figure 3.9: Cryo-EM data from the overfitting noise experiment All six picked sections on the left side were identified on a single micrograph. These images matched to the reference projection. The middle images show the aligned, picked non-particles. These images were reconstructed to a 3D representation of the protein complex. The images on the right side are the re-projections of the reconstructed map.

vulnerable to model-bias, the following experiment demonstrates that the classification of cryo-EM data is also model-biased.

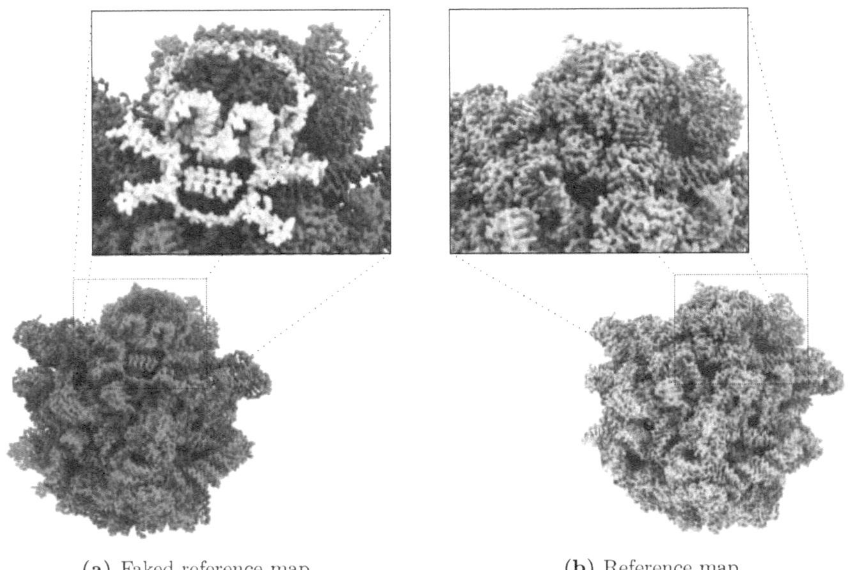

(a) Faked reference map

(b) Reference map

Figure 3.10: Reference models for the classification Here, the two reference models used to classify cryo-EM data are shown. The gray structure on the left side is the ideal representation of the protein complex. The teal model to the right was modified. This model has the identical structure as the gray map with an additional density shaped like a skull. The enlarged section are identical regions with respect to the map. Here, it is stressed that the density model of the skull is not part of the original protein complex structure.

Design of experiment A synthetic atomic model representing the structure of a skull (enlarged image in Figure 3.10) was created by Dr. Niels Fischer[2]. With *Chimera* a density map of the fake atomic model was built. This density map was added to a high-resolution refined structure map of the 70S ribosome. A set of $417,000$ single particle projection images with a box size of 420 and a pixel size of $0.75 \text{Å}/pix$ were classified without additionally aligning to the two references of the ribosome in Figure 3.10 with RELION. 80% of the projection images were assigned to the original reference in Figure 3.10. The other 20% showed a higher correlation to the faked-density reference in Figure 3.10. A gold-standard refinement was performed with the projection images which correlated better to the faked density. This computation was started using a reference structure that did not have the density of the skull to avoid reference biasing during alignment. It was possible because the data was already refined to an optimal structure and the imitation of the refinements was done with the known optimized parameter set. Important to notice is that there was no additional alignment procedure done. Here as well as before, the gold-standard refinement of the ribosome converged based on the correlation between the two maps.

Observation In Figure 3.11, the refined structure of one of the half maps and a meshed representation of the faked density model are shown. The refined 3D map of the ribosome contains a coarse density that fits to the skull reference. The enhanced side view in Figure 3.11b has a visible reconstructed density, where the atomic model of the skull was appended to the ribosome reference in Figure 3.10a. The top view of the map in Figure 3.11a, further, shows visible features of the silhouette of the faked density. The reconstructed density and the modeled density visually differ in their feature resolution. This, indeed, helps the reliability of the refined structure. To emphasize again, the skull is a faked atomic model, which was added to the ribosome map. The cryo-EM data was classified to this faked reference so that it was forced to detect a variation of the faked skull density in the recorded cryo-EM data. With the knowledge that the single particle projection images do not contain a signal similar to the skull structure, the classification fitted noise into the variation of the faked projected density. However, the classified cryo-EM was gold standard refined without the faked density reference in Figure 3.10a. The refinement should be free of the model bias based on the faked reference. Knowing the ribosome structure from previous experiments and the fake reference, the refined density of the skull is a contribution of the classified noise within the original single particle images. If the perfect density of the skull would have been reconstructed, it could be more obvious that there is something falsely detected in the cryo-EM data.

Both half maps have a representation of the reconstructed skull density such that the FSC in Figure 3.12 estimates a resolution of around 4.1 Å for the reconstruction with faked

---

[2]Department of Structural Dynamics, Max Planck Institute for Biophysical Chemistry

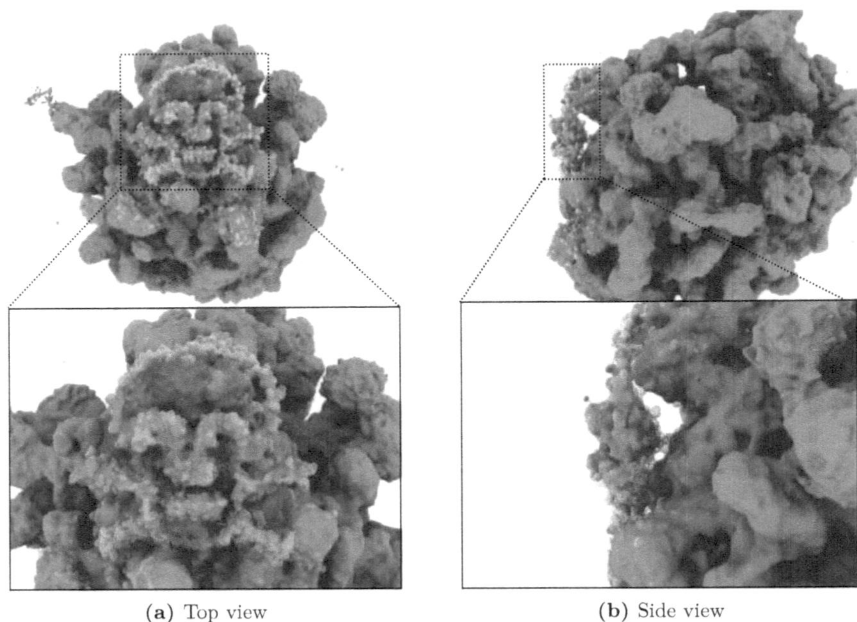

(a) Top view   (b) Side view

Figure 3.11: Refined structure of model-biased classified data The teal map in this figure is the reconstruction of the classified data with RELION. The two enlarged regions of the 3D reconstructed map (teal) show the recovered faked density in more detail. The meshed structure shaped as skull denotes a part of reference density, which was used to classify the data. The right image shows the top view of the reconstruction. The left image section shows a side-view of the 3D map.

density classified data (resp. 6.3 Å for the conservative threshold). Moreover, the FSC shows the general characteristics of the correlation curve between reconstructed proteins. Consequently, the FSC does not detect the mistakenly reconstructed noise and its resulting quality issues of the ribosome density. The estimated resolution of 4.1 Å is worse than the 2.9 Å published structure of the protein complex recovered from the identical cryo-EM data set [7]. One reason is the number of refined particles. Instead of refining 417k single particles only 83k projection images were reconstructed. Additionally, the resolution of the published map was estimated with the FSC between two auto-masked half maps. Using the auto-masking option in RELION the FSC for the experiment in Figure 3.12 determines a resolution of 3.1 Å (resp. conservative threshold 3.9 Å). The difference of 0.2Å is most likely a consequence of the number of projection images used to reconstruct.

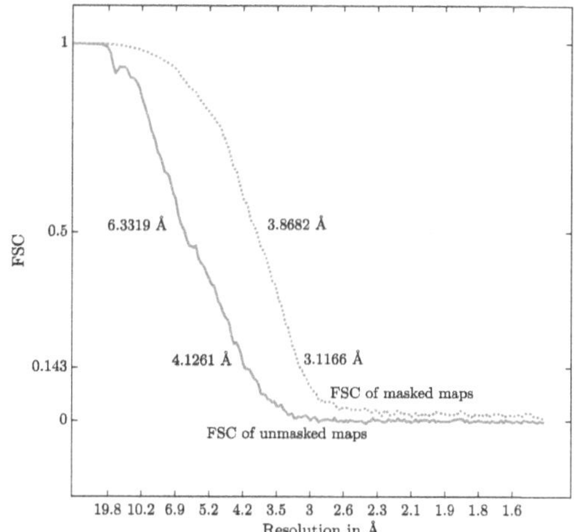

Figure 3.12: FSC of model-biased classified data Here, the unmasked and masked FSCs between the two half-set maps are plotted. The refinement was done with approximately 20% of the raw data, which classified well to the faked density reference map (the teal map in Figure 3.10).

3.2 Algorithm for validating and assessing the map quality

As seen in the previous experiments, several image-processing related errors can occur. In all three, the resolution of the structural maps was falsely detected and map errors remained undetected. With these experiments, it becomes more apparent that the FSC does not separate between signal and noise. Furthermore, the FSC is influenced by the reconstructed noise component. However, the advantage of these three experiments was that they were resulting in visually obvious erroneous structures, where the identification of the overestimated resolution is straightforward. Often the visual interpretation of the data is more difficult and published structure maps rely on the resolution measured by the FSC, which would have been misleading in all three experiments. The necessity to validate the quality of the reconstructed 3D model has been underlined by the previous experiments.

Once again, the FSC assess the resolution of the protein structure by correlating the two reconstructed maps. This correlation does not consider the relationship between the recorded signal and the reconstructed signal. In theory, the reconstruction of a protein complex should only recover signal which was detected and digitized into 2D single particle projection images. However during image processing, the noise of the projection images interferes so that false assumptions related to the protein complex structure are one of the consequences. The aim is to derive a validation tool linking the reconstructed signal of the map and the detected protein complex signal of the projection images and further, define the true resolution of the reconstruction based on their relationship.

Based on the cs-thm (see Theorem 2.2.6) the detected signal is a slice of the 3D Fourier object of the protein complex which in turn is equivalent to the 2D projection in Fourier space. The Fourier projection of the reconstructed map extracts a central slice with respect to a set of specific Euler angles. With this in mind the re-projection, which is the projection of the reconstructed map, and the projection image of the protein complex with respect to the identical degrees of freedom (see 1.2.1) are assumed to contain the identical signal information under the assumption of a perfect refinement. Generally, the re-projection either consists of the identical or less amount of signal as compared to the detected single particle projection. The measured distance between these two images is the residual of the reconstruction. Thereby, the SSNR could be defined as a ratio between the reconstructed signal and the residual. The SSNR is closely related to the FSC so that a resolution criteria based the ratio between the reconstructed signal and the residual of the detected and reconstructed signal can be established.

3.2.1   Algorithm basics

The single particle projection image $I_r^{\psi_k}$ was identified on the micrograph recorded with a TEM. It was optimized with respect to the translation and rotation parameters of the image. Ideally, this image, which is used during the validation step, was not altered by algorithms and digitally overwritten to prevent bias or algorithmic errors. The second image is the projection of the reconstructed protein complex I_s^{ψ}. This projection depends on the angle set ψ of the single particle image I_r^{ψ}.

Let $I_r^{\psi_k}$ be a detected single particle image, where $k \in N$ is the k-th image in the total of N images within the data set. The projection image $I_s^{\psi_k}$ is the projection of the refined map with respect to $\psi = (\alpha, \beta, \gamma)$. ψ defines the optimized Euler angles for the k-th image. As defined in section 2.2 the raw projection images are specified by the detected signal and an additive noise component M with Gaussian distribution of zero-mean and variance equal to one.

$$I_r^{\psi_k} = F + M_k \tag{3.2}$$

$$I_s^{\psi_k} = F_s, \tag{3.3}$$

where F_S is the reconstructed signal in Fourier space. Each I_r^{ψ} has a signal component and an independent noise component M_k. Therefore, the sum over multiple I_r leads to a decrease in noise. The aim is to introduce a validation strategy which relates the reconstructed signal with the detected signal. Hence, the euclidean distance between these two images is computed.

$$I_r^{\psi_k} - I_s^{\psi_k} = (F + M_k) - F_s \tag{3.4}$$

This distance (3.4) has two possible outcomes

$$
= \begin{cases} M_k & F = F_s \\ M_k + F_\delta & \text{else} \end{cases},
\tag{3.5}
$$

where F_δ describes any deviation of the signal components. On the one hand the refinement is able to completely recover the signal of interest. On the other hand the signal of interest was not fully reconstructed. If this occurs, there is a residual signal F_δ. This unrecoverable signal is considered to be noise.

3.2.2 Algorithm

The validation approach introduced in the following section is computed in Fourier space. Therefore, all images are Fourier transformed (see subsection 2.2.3). At first the signal and the noise components with respect to the Fourier rings, e.g. in Figure A.1, are computed.

$$
\Gamma_k^S(r, \Delta r) = \sum_{R \in (r, \Delta r)} \left| I_s^{\psi_k}(R) \right|^2,
\tag{3.6}
$$

where $\Gamma_k^S(r, \Delta r)$ is power spectrum of the k-th re-projection signal with respect to its Fourier rings.

$$
\Gamma_k^N(r, \Delta r) = \sum_{R \in (r, \Delta r)} \left| I_r^{\psi_k}(R) - I_s^{\psi_k}(R) \right|^2,
\tag{3.7}
$$

where $\Gamma_k^N(r, \Delta r)$ is the power spectrum of the k-th distance between the re-projection and the raw data with respect to its Fourier rings. The residual of the distance should represent the noise in the images. Similar as explained in (3.5) there is only the noise or unexplained signal in the difference. The Equation 3.7 is the distance as defined by (3.4) between the extracted single particle image and the projection image of the refined structural map. The signal explains all the refined signal. The noise is the difference between the noisy signal outcome of the TEM and the reconstructed signal. In the best case the noise is the truly detected noise. The worst case gives a residual of an unexplained signal additionally to the signal. Both equations lead to the following fraction called Quality Signal-to-Noise-Ratio (QSNR).

$$
QSNR_k^S(r, \Delta r) = \frac{\Gamma_k^S(r, \Delta r)}{\Gamma_k^N(r, \Delta r)},
\tag{3.8}
$$

where $QSNR_k^S(r, \Delta r)$ is the rotationally averaged spectral variance ratio based on the signal being reconstructed and the noise being the unrecoverable signal as well as the disturbances of the predictions. The $QSNR_k^S(r, \Delta r)$ is computed with respect to the k-th image in the data.

Noise reduction   During the refinement the data is averaged over the spatial frequencies where the central slices intersect based on the cs-thm. That leads to a noise reduction in the spatial frequencies additionally depending on the number of averaged images. Because of this it is difficult to measure a ratio between the noise-reduced re-projected data and the residual of the refinement.

One of the main drawbacks to theoretically determine the attenuation factor results from the various number of input options for the refinement. Hence, Unser *et al.* [76] introduced an empirical model to find a factor incorporating the amount of reduction for each spatial frequency. A ratio between an artificial noise AN image and the re-projection image RN of an artificial noise reconstruction. It means that the artificial noise AN is refined with the identical parameter as the raw cryo-EM data was refined. In this **book** we followed the idea presented by Unser *et al.* [76].

$$\Gamma_k^{RN}(r, \Delta r) = \sum_{R \in (r, \Delta r)} |RN_k(R)|^2 \,, \tag{3.9}$$

where $\Gamma_k^{RN}(r, \Delta r)$ is the power spectrum of the k-th re-projection of the reconstructed noise with respect to its Fourier rings.

$$\Gamma_k^{AN}(r, \Delta r) = \sum_{R \in (r, \Delta r)} |AN_k(R)|^2 \,, \tag{3.10}$$

Again, this leads to a quotient between the reconstructed noise and the artificial input noise component.

$$QSNR_k^N(r, \Delta r) = \frac{\Gamma_k^{RN}(r, \Delta r)}{\Gamma_k^{AN}(r, \Delta r)} \tag{3.11}$$

where $QSNR_i^N(r, \Delta r)$ is the rotationally averaged spectral variance ratio of the noise reduction based on the optimized signal parameter. The quotient is computed with respect to the k-th image within the data set. In all, this provides a factor how much noise was eliminated through averaging over the recorded projection images $I_r^{\psi_k}$.

QSSNR   Combining both equations, (3.8) and (3.11), the QSSNR is the ratio between the power of reconstructed signal and the residual with respect to the measured signal. Here, the QSSNR is a function of spatial frequency with respect to Fourier rings $(r, \Delta r)$ of the images.

$$QSSNR(r, \Delta r) = \frac{1}{N} \sum_{k=1}^{N} \frac{QSNR_k^S(r, \Delta r)}{QSNR_k^N(r, \Delta r)}, \tag{3.12}$$

where the $QSSNR(r, \Delta r)$ is the sum over all signal-residual-ratios (Equation 2.46). To emphasize again, the $QSNR^S(r, \Delta r)$ is the ratio between the reconstructed signal over the whole image set and the noise over all refined images as defined in (3.8) and the $QSNR^N(r, \Delta r)$ is the factor of noise reduction of all refined images as defined in (3.11). An unbiased estimate of the SSNR was determined by Unser *et al.* [72].

$$S(r, \Delta r) = max(0, QSSNR(r, \Delta r) - 1) \tag{3.13}$$

The relationship between the FSC and SSNR explained in subsection 2.5.2 should also hold true for the QSSNR. As a consequence, a FRC of projections, based on the QSSNR, should be computed with

$$\text{FRC of projections}(r, \Delta r) = \frac{S(r, \Delta r)}{S(r, \Delta r) + 2} \tag{3.14}$$

The FRC of projections should ideally result in a more reliable resolution determination. It is an estimator based on the ratio between the reconstructed signal and its residual to the recorded cryo-EM data.

3.2.3 Implementation

The validation approach described in the previous section was implemented in MATLAB 2018a and MATLAB 2017b. Sigworth [77] has published read and write functions written in MATLAB specific for cryo-EM data files. Functions such as ReadMRC, WriteMRC as well as WriteMRCHeader are published in a repository [77]. The implementation of the validation approach made use of these scripts. Built-in functions implemented in MATLAB Fast Fourier Transformation were also resources used for the implementation here. The two main components are shown in the following excerpts of the source code. The $QSNR^S$ is computed by looping over all images within the data set (see Listing 3.1) and computing the power spectra of the reconstructed signal as well as the residual. Afterwards, the sum over all elements in the rings is computed.

```matlab
% esitmate the SNR of the reconstructed signal to the residual between
    the reconstruction and the recorded data
for i = 1:N
estSignal              = abs(I2(:,:,i)).^2;
rSignal                = ring .* estSignal;
sumSignal              = squeeze(sum(sum(rSignal,2),1));
weightSignal_R         = 1./nr .* sumSignal;
```

```matlab
Residual              = abs(I3(:,:,i)).^2;
estResidual           = sum(Residual,3);
rResidual             = ring .* estResidual;
sumResidual           = squeeze(sum(sum(rResidual,2),1));
weightResidual_R      = 1./nr .* sumResidual;

SNR(i,:)              = weightSignal_R ./ weightResidual_R;
end
```

Listing 3.1: $QSNR^S$ for each image

The fraction $\frac{1}{n_r}$ denotes the number of elements in the current Fourier ring as the number of elements increases with increasing radius (Figure A.1). This is beneficial to normalize the number the elements in the rings with increasing radius since these Fourier rings accumulate more spatial frequencies. The noise reduction ratio in Listing 3.2 is computed equivalent to the QSNR as seen in Listing 3.1.

```matlab
% esitmate the SNR of the reconstructed noise to the noise
for i = 1:N
estSignal             = abs(I2(:,:,i)).^2;
rSignal               = ring .* estSignal;
sumSignal             = squeeze(sum(sum(rSignal,2),1));
weightSignal_R        = 1./nr .* sumSignal;

estANoise             = abs(I3(:,:,i)).^2;
rANoise               = ring .* estANoise;
sumANoise             = squeeze(sum(sum(rANoise,2),1));
weightANoise_R        = 1./nr .* sumANoise;

SNR(i,:)              = weightSignal_R ./ weightANoise_R;
end
```

Listing 3.2: $QSNR^N$ for each image

At last the QSSNR in (3.12) is computed. Assuming gold-standard refined cryo-EM data the relation between the FRC of projections in (3.14) is computed with respect to the QSSNR of the half-sets (see subsection 2.5.2).

```matlab
% calculate the QSSNR of the reconstruction set
QSSNR   = max(0,1/N * SSNR./SNR - 1);
FSCP    = QSSNR ./ (QSSNR + 2);
```

Listing 3.3: Estimating the QSSNR and the FRC of projections

3.3 Application of the validation algorithm

The aim of the presented algorithm is to detect overestimated resolution claimed by the FSC. Further, it aims to validate the estimated resolution of the cryo-EM structure maps based on the recorded signals. The noise present in real cryo-EM data interferes with the data analysis as presented in 3.1. The noise is also difficult to quantify. The consequence is that tests, which are carried out with cryo-EM data are not reliable to verify the approach. Tests with synthetic data sets were carried out. However, as the noise is difficult to define the artificial noise in the synthetic data would most likely not behave like cryo-EM noise. Consequently, a test with correct identified resolution data as well as overestimated resolution data is executed. Especially, the false detected resolution data, which was presented in section 3.1, is suited to demonstrate the quality of the validation approach. A short reminder, the resolution of a map is described by the structural features of the protein complex, which correspond to a specific spatial frequency (see Equation 1.5).

In the following, the resolution curves determined by the validation algorithm and the refinement are compared. To distinguish between these two curves the **book** introduces the two distinct names. The FRC of projections (see (3.14)) is the correlation between two objects based on the introduced validation approach (section 3.2). The Fourier Shell Correlation of reconstruction (FSC of reconstruction) is the FSC curve determined between the two reconstructed (resp. refined) volumes. The FSC of reconstruction is taken from the software package which was used to compute the structures. To compare the estimated resolution of the structures the FSC thresholds of 0.143 and 0.5 (as described in section 3.2) are considered. The QSSNR with respect to the data set is also shown. The threshold for the QSSNR is one. At this level the reconstructed signal and the residual have the same power. Everything below one implies that more noise and wrongly reconstructed signal is present.

Artificial noise The noise reduction during the reconstruction is not negligible for a method comparing re-projections and projection images. In the derivation of the validation algorithm reconstructing noise with the identical optimized projection image parameters was introduced (see section 3.2.2). For the application of the algorithm, an artificial ran-dom noise set was generated with MATLAB. The software has a built-in function called white Gaussian noise wgn(), which generates random noise data with zero-mean and vari-ance of one. A second possibility was to extract noise from the recorded micrographs as mentioned in subsection 3.1.2. The artificial noise images were reconstructed with the identical parameter set in the same software package.

3.3.1   Synthetic data

The algorithm derived in subsection 3.2.2 is supposed to verify the feature resolution of the reconstructed structures. To validate the quality of the approach synthetic test data is used since the correctness of experimental data is unknown. Due to various reasons, e.g. the low SNR, the protein complexes reconstructed from cryo-EM data resolve to different nominal feature resolutions. Hence, the validation approach needs to be able to detect different resolutions (see section 1.4). In this **book** two tests with differently resolved synthetic data sets were executed to validate the derived algorithm. The structure of the ribosome used in subsection 3.1.2 is a sufficiently resolved model for both test designs. Due to box size related issues such as computation time a smaller component of the ribosome, namely the 50S ribosomal protein L13 (see Figure 3.14) was extracted. The protein has a box size of $104 \times 104 \times 104$ and voxel size of 0.75 Å. Prior to processing the protein, a total of 6144 random noise images were created with MATLAB. Here, the artificial noise images are independently Gaussian distributed with zero-mean and a variance of one. These images were added onto the projection images of the protein. To ensure uncorrelated additive noise the 2D correlation between all noise images was calculated. The highest correlation between two noise images within the data set is 0.05 (see Figure 3.13).

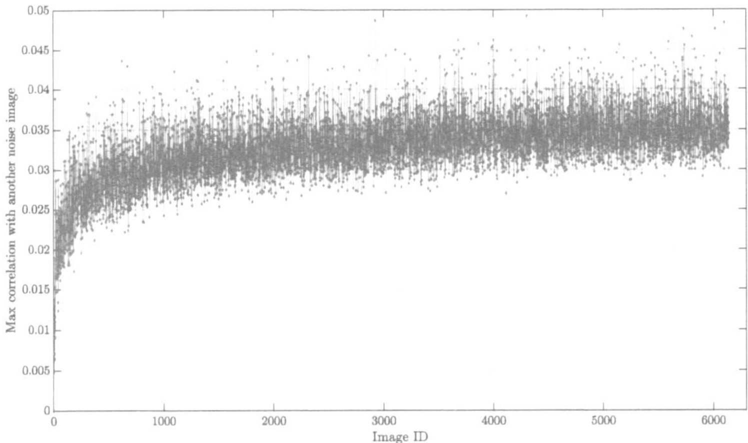

Figure 3.13: Maximum correlation between artificial noise images Here, the highest correlation between two images within the set of artificial noise images is plotted. On the x-axis the index of the image number is given. On the y-axis the maximum correlation of the i-th image with respect to all images in the data set is given.

Design of experiment   In the first test the 50S ribosomal protein L13 with high resolution features in Figure 3.14 was used. The second test was done with a Gaussian filtered map of the 50S ribosomal protein L13 (see Figure 3.14). Fourier filtering the density map removed all spatial frequencies higher than a specified threshold (see 2.4.1). The filtered 50S ribosomal protein L13 in Figure 3.14 has only features up to a resolution between 1.9 Å and 2 Å (see Figure 1.11). Both maps are representations of protein structures at high resolution since the noise in cryo-EM influences especially the detection of high resolution features.

Both maps were Fourier projected by an angular distance of 3.66 with *CowEyes*. The high variation of the pixels in the images forced a standardization of the data to zero-mean and variance of one. The first 3072 of the 6144 random noise images were added onto these projections (see subsection 2.3.2). At this point the projection images contained an equal amount of signal power as well as noise power. The 3072 noisy protein images were randomly split into two subsets of 1536 projection images and then independently reconstructed with *CowEyes*. The reconstruction was a straightforward Fourier reconstruction due to the known Euler angles. The set of the remaining 3072 artificial, random noise images was reconstructed based on the 50S ribosomal protein L13 reconstruction parameter set with *CowEyes*. These noise images had a maximal correlation of 0.05 so that the corresponding estimation of the resolution should most likely result from the correlation of the synthetic data.

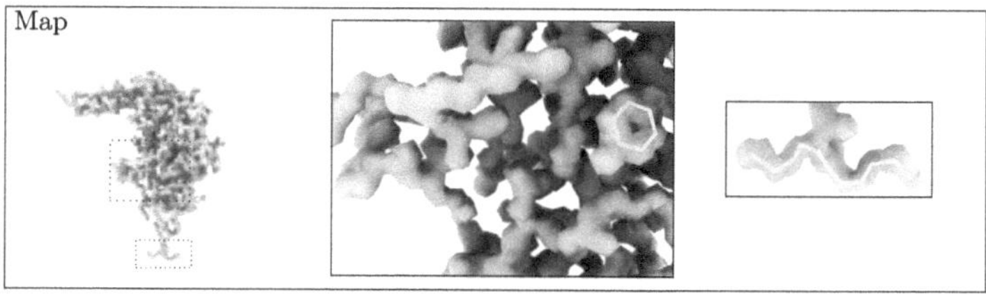

(a) Density map of the 50S ribosomal protein L13

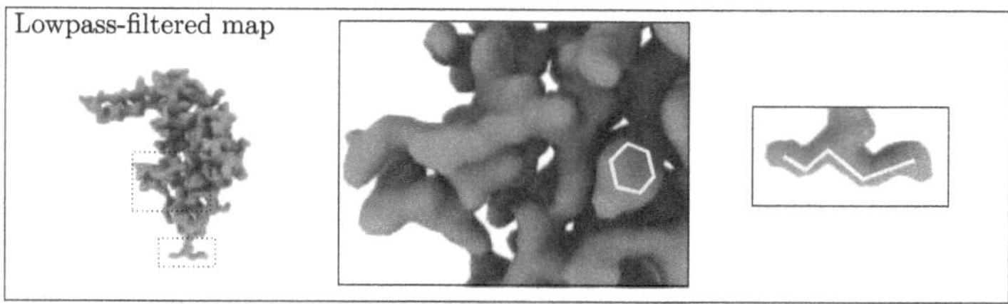

(b) Density map of the 50S ribosomal protein L13, which was Gaussian lowpass-filtered. That means that all spatial frequencies higher than 0.5128 are removed.

Figure 3.14: Synthetic test map The 50S ribosomal protein L13 is part of the ribosome. Both maps were created from the atomic model of the protein in Figure 3.7. The highlighted hexagons in both enhanced regions represent a chemical feature of the protein, the hexagonal benzene group of the tyrosine residue. The map in Figure 3.14a has a detailed representation of the geometrical shape so that it is possible to see the ring property. In contrast, the lowpass-filtered map in Figure 3.14b shows a less resolved density around the identical area of the hexagon. The second enhanced region of the protein underlines the difference between the two feature resolutions.

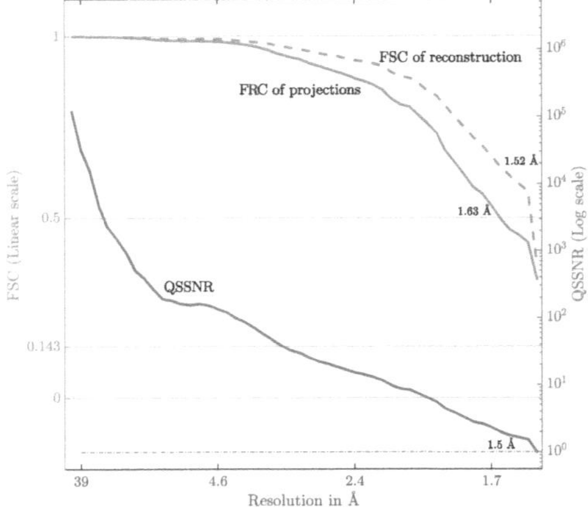

Figure 3.15: Comparing the FSC of reconstruction and FRC of projections for synthetic data Here, the QSSNR has an equal amount of reconstructed signal power and residual power until 1.5Å. The resulting FRC of projections determines the resolution at the same spatial frequency with 1.63Å for the threshold of 0.143. The FSC of reconstruction computed with *CowEyes* between the two reconstructed half-maps states the feature resolution at 1.52Å.

Observation In Figure 3.15, the FSC of reconstruction which was computed between the two half-set maps of the ideal protein (see Figure 3.14a) is shown. The FSC of reconstruction does not drop below the resolution threshold 0.143. The conclusion is that the structure resolved until the Nyquist frequency. If the more conservative threshold of 0.5 is applied, the feature resolution of the reconstructed map is 1.52 Å. Based on the FSC of reconstruction between both half-maps the reconstructed protein complex resembles a near-perfect copy of the synthetic input data. Further, all spatial frequencies of the underlying signal were recovered during the reconstruction. Indeed, the data of this test has ideal signal, which was only disturbed by a synthetic Gaussian distributed noise of zero-mean and variance of one. Besides the perfect noise model, the optimal reconstruction parameters are known for each image. The design of the experiment was engineered to result in a good resemblance of the input data. The resolution estimated by the FSC of reconstruction supports the expected behavior. Further, the QSSNR, which is the ratio between the reconstructed signal and the residual of the reconstruction (see section 3.2.2), also estimates a resolution of 1.5 Å at the threshold of one. The validation approach introduced a FRC of projections based on the QSSNR. The FRC of projections estimates the resolution around 1.63 Å at 0.5. The FRC of projections descends further such that the feature resolution of the reconstructed data would also be estimated at 1.5 Å. Both FSC curves provide the conclusion that the resolution of the protein structure is high. Both correlation curves, the FRC of projections and the FSC of reconstruction, also show similar curve characteristics. However, the FRC of projections is slightly below the FSC of reconstruction, especially for the higher spatial frequencies. A possible interpretation is that the FRC of projections is able to recover a difference between the ideal and the reconstructed signal. It is a reasonable assumption because the synthetic data was still disturbed by the identical power of noise as signal present. On that note, it is possible to assume that the resolution is a validation of the reconstructed signal since the FRC of projections was defined as the resolution based on the ideal and the reconstructed signal. Finally, this could be an indicator of well measured resolution based on the FRC of projections.

A single test is not a sufficient verification of the validation approach (see subsection 3.2.2). The second test was done with the less resolved model in Figure 3.14b. Once again, the FSC of reconstruction (see Figure 3.16) was computed between the two reconstructed half-set maps. The resolution of this structure was estimated to be 2.02Å at 0.143. It is known that the input map was Fourier filtered to 2 Å before the projection. The highest possible reconstructed feature corresponded to 2 Å. Because of this, the reconstruction of the synthetic data is assumed to be an optimal replica of the protein.

The validation algorithm was also applied to this data. The QSSNR in Figure 3.16 (defined in section 3.2.2) estimates the resolution to be 2.55 Å at the threshold of one. The corresponding FRC of projections drops below 0.143 at 2.16Å. Here, the predicted resolution of the reconstructed map differs by 0.14 Å. Here too, both FSC curves seem to come to the similar general conclusion about the feature resolution of the reconstructed protein. Once again, the FRC of projections shows the similar decreasing behavior of the FSC of reconstruction and is slightly below the FSC of reconstruction. Analogous to the conclusion drawn above, the FRC of projections could predict a more reasonable resolution of the reconstructed data. Here too, the reconstruction parameters were optimal and the power of the noise was identical to the power of the signal. However, noise is a random disturbance such that the reconstruction could be influenced by this. The predicted resolution by the FRC of projections is a valid estimate.

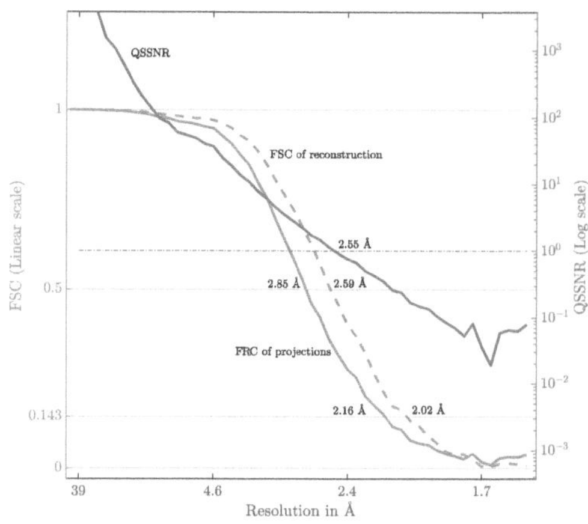

Figure 3.16: Comparing the FSC of reconstruction and FRC of projections for synthetic data Here, the FSC of reconstruction computed with *CowEyes* between the two reconstructed half-maps states the feature resolution at 2.02 Å for the threshold of 0.143. The FRC of projections between the recorded non-single particles and the re-projections determines the resolution at 2.16 Å at the same threshold. The QSSNR has an equal amount of reconstructed signal power and residual power at 2.55 Å.

3.3.2 Experimental data

The noise in cryo-EM projection images is difficult to quantify. Because of this, modeling synthetic noise, which behaves like the experimental detected noise, is complicated. As a consequence, the synthetic data does not resemble the true behavior of noisy cryo-EM data. The tests which are described above are only indicators of the validity of the implemented approach (see subsection 3.2.2). Therefore, it is important to execute tests with experimental cryo-EM data. The **book** provides three tests, which are based on either an accurately estimated resolution or overestimated feature resolutions for cryo-EM density maps and the recorded data. The aim of a test with a reliable estimated resolution cryo-EM density map is to establish that the algorithm in subsection 3.2.2 is capable of confirming the resolution. Here, more reliable means that the estimated resolution of the 3D cryo-EM maps fits the resolved features. The cryo-EM density map was visually evaluated to its chemical correctness. Consequently, both correlation curves, the FSC of reconstruction and FRC of projections, are expected to come to a similar conclusion about the resolution of the 3D maps.

Further on, the proposed method was introduced to validate whether the reconstruction is a true reflection of the recorded data or, e.g., overfitted noise. Oftentimes overfitting results in an overestimated resolution of the cryo-EM density map. To verify the implemented approach, falsely estimated experimental resolution data is tested. One experiment in section 3.1 is well suited for this test. The overfitting of noise as described in subsection 3.1.2 provides a transparent and straightforward test to verify whether the implemented approach validates the resolution of the reconstructed cryo-EM data or not. The false interpretation of the resolution is not debatable since the recorded projection images (Figure 3.18a) do not contain any signal related to a protein complex. Additionally, the difference between the detected signal and the reconstructed signal (Figure 3.18b) is visible and cannot be further questioned. This is an obvious overestimation of the resolution. Because of this a second test examines the faked classified cryo-EM data described in subsection 3.1.3. Here, the validation approach should rather detect the quality problem of the 3D reconstructed map. The introduced classification bias results in a fine structural detail within the reconstructed cryo-EM map. The FSC of reconstruction does not detect the qualitative issues. Ideally, the QSSNR determines the difference in the two signals and further results in a worse FRC of projections.

3.3.2.1 T20S proteasome

Design of experiment   This experiment aims to verify the validation approach for true estimated resolution data. In subsection 3.1.1, high-resolution cryo-EM data was introduced. With an accurate CTF correction a subset of the T20S proteasome refined up to 2.8 Å. This data was visually assessed and evaluated to have a valid estimated resolution.

An optimal set of 115,000 T20S proteasome projection images which were gold-standard refined to high resolution was evaluated in the following experiment. These single particle images were reconstructed with RELION. The FSC, computed with RELION and here called FSC of reconstruction, determined the feature resolution of the protein complex map. The reconstructed map of the T20S proteasome was projected with RELION based on the refined optimal parameter. To account for the noise reduction during the refinement the artificial noise was also reconstructed and further projected with the identical refined parameter set. Finally, the re-projection images of the protein complex and of the reconstructed noise as well as the recorded projection images and the artificial noise were processed with MATLAB (see subsection 3.2.3). In MATLAB, the QSSNR and FRC of projections were computed (Appendix B).

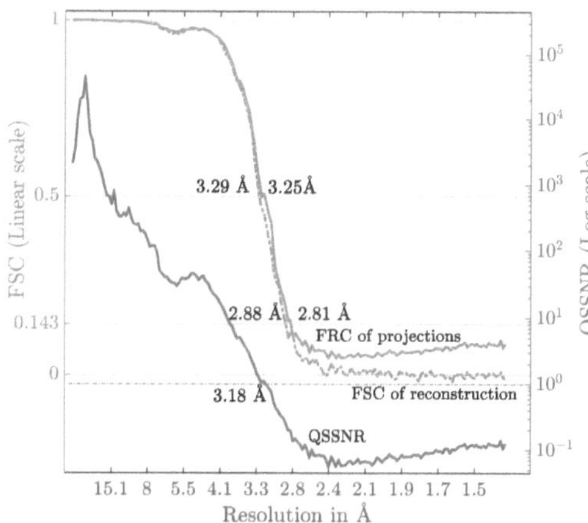

Figure 3.17: Comparing the FSC of reconstruction and FRC of projections for the experimental T20S proteasome Here, the FSC of reconstruction computed between the two T20S proteasome half-maps with RELION detects the resolution at 2.88 Å. The FRC of projections between the original particles and the re-projections is shown and determines the resolution at 2.81 Å. The QSSNR (see 3.2.2) has an equal amount of reconstructed signal power and residual power at 3.18 Å.

Observation In Figure 3.17, the FSC of reconstruction and FRC of projections are compared. Based on the gold-standard refinement RELION determined the resolution at the level of 0.143. At this point the two half-maps coincide up to a spatial frequency of 0.35, which corresponds to the feature resolution of 2.81 Å. The FSC of reconstruction was a correct interpretation of the true resolution of the map. The QSSNR estimates the feature resolution 3.18 Å for the data. The FRC of projections is computed from the QSSNR. At 0.143 the FRC of projections determines the resolution of 2.81 Å. As it is often criticized that 0.143 is too optimistic, the second cut-off level considered in the book is 0.5. At this point both, the FSC of reconstruction and FRC of projections, determine the resolution of about 3.2 Å. Both approaches come to the same feature resolution for the reconstructed maps.

The high-resolution T20S proteasome structure visually contained the geometrical features (see section 1.4). The FSC of reconstruction resulted in a reasonable definition of the resolution. The FRC of projections also estimated a similar resolution of the protein complex map. Both, FSC of reconstruction and FRC of projections, coincide for the most part. The difference of 0.07 Å between the two correlation measurements at 0.143 could be result from better resolution estimation based on the recorded data. However, it could also imply that noise information was measured. The curve characteristics equal until 2.6 Å. A possible conclusion is that the recorded signal in these spatial frequencies is dominated by the noise and hence, not a confident estimation. The noise directly influences the denominator of the QSSNR and hence, it impacts the FRC of projections. After all, the FRC of projections could be a valid measurement for the underlying true resolution.

3.3.2.2 Fitting noise

Design of experiment This experiment aims to determine if the validation algorithm is able to detect an unmistakably overestimated resolution. The ribosome was reconstructed by fitting pure noise into a map of the protein complex (see subsection 3.1.2). The FSC of reconstruction in Figure 3.19 defined a feature resolution of 5.6 Å. The non-particle projection images in Figure 3.18a show the real recorded data. There is visually no signal of the ribosome present. Even with the assumption of a low SNR, it is known that only grids with a thin carbon support film were imaged. In contrast, the re-projection images of the reconstructed map show a reconstructed ribosome signal in Figure 3.18b. There exists a visual difference between these two sets of images. The idea of the validation algorithm was to find the distance between these images and define it as the residual of the reconstruction. In the best case, the QSSNR and the FRC of projections result in a resolution of infinity. The resolution of infinity could be interpreted as that there is no relationship between the recorded and reconstructed data. The processing of the data was described in subsection 3.1.2. Once again, the artificial noise images were constructed with MATLAB and equivalently processed as the non-particle projection images in Figure 3.18a.

Observation The FSC of reconstruction implies a feature resolution of 5.6 Å at 0.143. In Figure 3.19, the computed QSSNR drops below the threshold of one at the spatial frequency of 0.1157. Consequently, it would be assumed that the reconstructed signal is valid up to 8.64 Å. This is contradicting to the knowledge that the data does not contain a protein complex signal. Furthermore, the FRC of projections estimates a resolution of around 3.19 Å at 0.143. This is an even higher estimate than the FSC of reconstruction. The cryo-EM micrographs were template picked which knowingly introduces a model bias. Consequently, it could be assumed that a more conservative threshold should be applied. However, both correlation measurements determine a resolution of the faked ribosome map of around 9 Å at 0.5. The more conservative threshold for the FSCs also fails. All estimations of the resolution contradict with the prior knowledge about the recorded cryo-EM

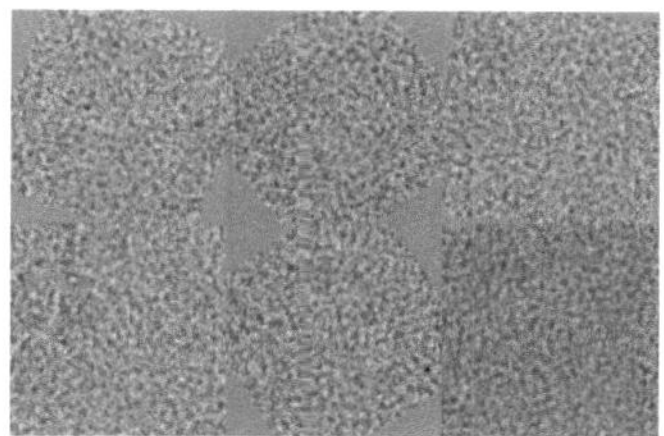

(a) The identified projection images of the protein complex based on a template picking are shown. The particles do not contain a real signal from a protein complex as described in subsection 3.1.2.

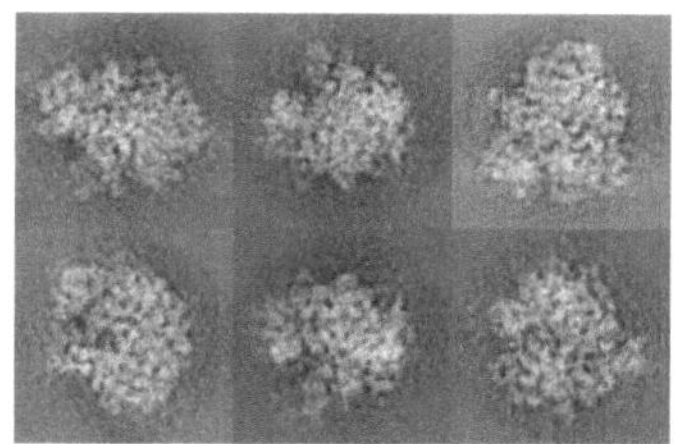

(b) The re-projections of the reconstruction based on pure noise images are shown. The re-projection corresponds to the same degrees of freedom as determined for the original non-particle images.

Figure 3.18: Non-particle projection and re-projection image

data.

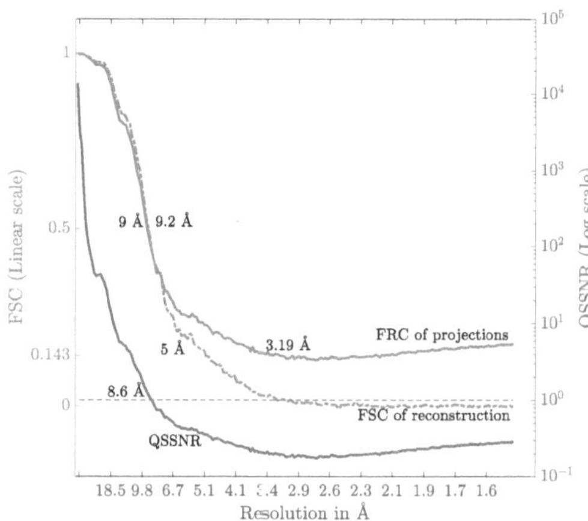

Figure 3.19: Comparing the FSC of reconstruction and FRC of projections for the experimental noise fitted data Here, the FSC of reconstruction computed with *CowEyes* between the two reconstructed half-maps states the feature resolution at 5.6 Å. The FRC of projections between the recorded non-single particles and the re-projections determines the resolution at 3.19 Å for the threshold of 0.143. The QSSNR has an equal amount of reconstructed signal power and residual power at 8.64 Å.

The algorithm fails to detect the quality of the data. The QSSNR and FRC of projections cannot distinguish between the pure noise and the faked reconstructed signal. In theory, the power of the residual in Equation 3.7 and the power of the reconstructed signal in Equation 3.6 should be equal from the first spatial frequencies. The expected behavior would be that the QSSNR drops below the threshold of one within the first or second spatial frequency. These frequencies often correspond to nominal resolution numbers higher than a reasonable estimate would be (more than 100 Å). Nevertheless, the algorithm did not come to this conclusion. Consequently, the introduced algorithm is also no validation approach for noisy cryo-EM data.

3.3.2.3   Classification of noise

Design of experiment The classified data presented in subsection 3.1.3 was used to test whether the validation algorithm can detect a detailed quality issue within the protein complex structure. Ideally, the distance between the overfitted noisy parts of the single projection images and the corresponding re-projections of the reconstructed 3D protein complex map in Figure 3.20a should show a discrepancy between the signals. Consequently, the QSSNR should measure a higher residual power and decrease faster. The processing of the data was described in subsection 3.1.3.

To validate the data the amount of noise reduction needs to be determined. Therefore, the artificial noise (see Figure 3.20b) was refined with the identical parameter set with RELION. Important to notice is that the noise refinement was initiated from the optimal high resolved ribosome data. The ribosome model which was used as a reference to refine the ribosome data in RELION was not changed and thus, present during the noise refinement. After the refinement the reconstructed ribosome in Figure 3.11 and the noise volume are projected. The four image stacks were further processed with MATLAB. The QSSNR and the FRC of projections are computed.

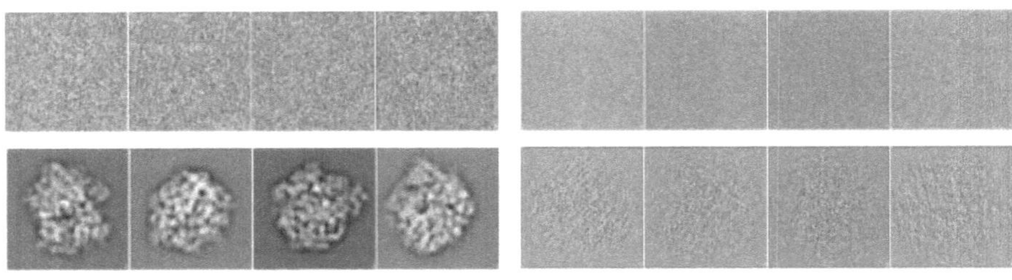

(a) Here, the projection images of the ribosome and the re-projections of the reconstructed structure are shown.

(b) Here, the artificial noise images and its re-projections are shown. The re-projections show a model bias with respect to the ribosome reference after refining the artificial noise.

Figure 3.20: Classified cryo-EM data

Observation The FSC of reconstruction determined a resolution of the maps up to 4.097 Å. The QSSNR decreases in the first spatial frequencies. However, it starts to increase again between the spatial frequencies related to 7 Å and 5 Å. It never drops below the threshold of one. As a consequence, the feature resolution could be assumed to infinity. Depending on the interpretation it could either be a structure resolved up to atomic resolution or a structure biased through image processing. Analogous to the QSSNR, the FRC of projections cannot define a feature resolution. The validation approach fails to detect the resolution. Furthermore, both curves, the QSSNR and the FRC of projections, are contradictory to their expected decreasing behavior.

The unexpected curve characteristics could result from the determination type of the noise reduction. In comparison to the test in subsubsection 3.3.2.1 the cryo-EM density maps are direct results of the refinement. Therefore, the noise reduction was done by refining the artificial noise with RELION. The reference model which was used to refine the artificial noise was a detailed representation of the ribosome. This could have initiated the overfitting of the noise. The re-projection images in Figure 3.20b show a slight influence of the model to the noise. The centers of the re-projected noise volumes show a different performance than the background with respect to the noise reduction.

A noise reduction without a model bias could be determined by starting the RELION with an artificial noise volume. However, the refinement of the ribosome data was initiated after is had been reconstructed to a high resolution. Consequently, the refinement of the falsely classified data was started with the last iterated density map of the original refinement. To imitate the original set up the reference model was not changed.

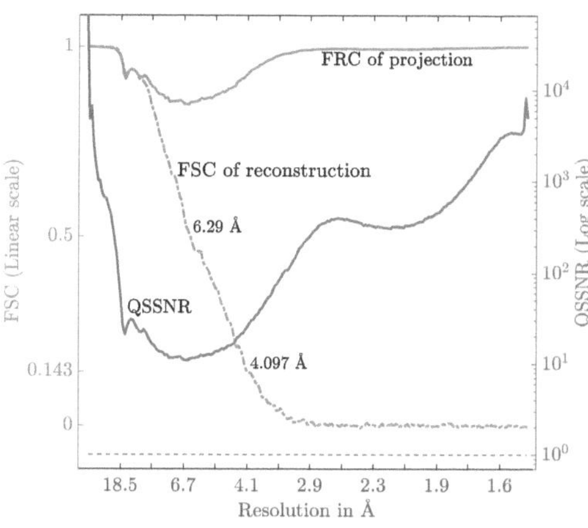

Figure 3.21: **Comparing the FSC of reconstruction and FRC of projections for the experimental classified noise** The FSC of reconstruction estimates the resolution of the reconstructed ribosome of 4.1 Å. The QSSNR and FRC of projections never fall below the threshold which specifies the resolution. Further, both increase in the high spatial frequencies.

If a different noise reduction estimation, e.g. only a reconstruction, would be applied, the QSSNR and the FRC of projections could come to a different conclusion about the feature resolution. However, the FRC of projections most likely would still fail to detect the qualitative difference between the recorded and the reconstructed signal since the structural dissimilarity of the two signals, here, is too detailed. That follows from the experiment done with the non-particle picked data in subsubsection 3.3.2.2, where the FRC of projections also failed to validate the signal.

3.4 Investigation of the validation algorithm

The tests of the experimental overestimated resolution data (subsubsection 3.3.2.2 and subsubsection 3.3.2.3) demonstrated that the introduced approach also fails to estimate the true resolution of the data. The test with correctly estimated resolution data set (see subsubsection 3.3.2.1) does also not indicate whether the introduced validation approach estimated the true resolution or measured noise. It could be that the distance in Figure 3.22d between the single particle image and the re-projection of the structure does not resemble the true difference between the two signals. It could follow from the normalization of the recorded data. Besides this, the residual signal might not have been a true distance between the recorded and the reconstructed signal, but a deduction of noise. Because of this, the noise could dominate the distance. Thus, in the next part of the **book** the underlying issues are discussed.

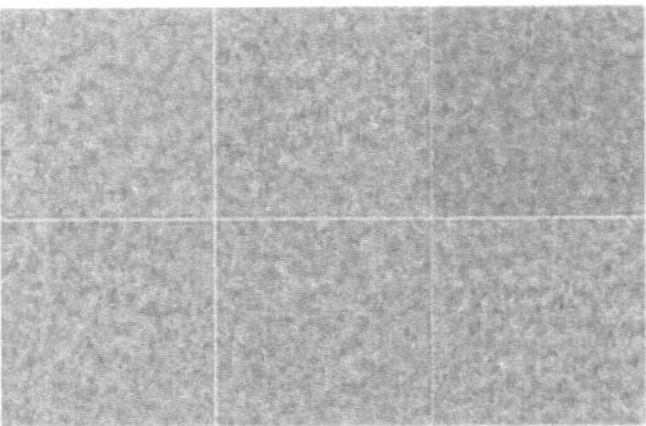

(a) Normalized detected T20S proteasome projection images. The images have a low SNR. The mean only differs from zero from the third decimal digit and the variance is close to one. The gray values vary from -5.5 to 4.68.

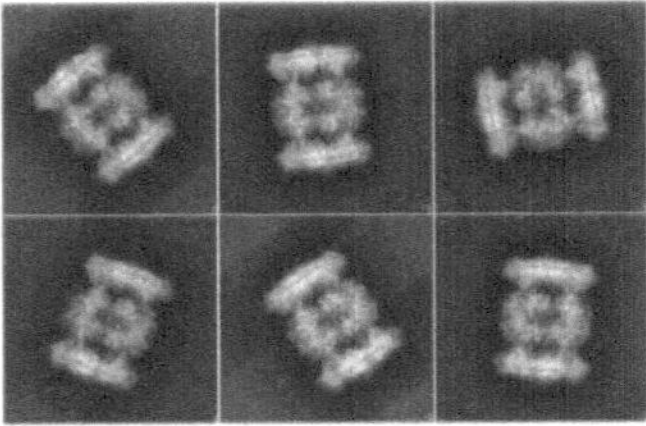

(b) Re-projections of the reconstructed T20S proteasome. The mean is around -0.03 and the variance is close to zero $(0.08-0.09)$. The gray values vary from -0.44 to 1.58.

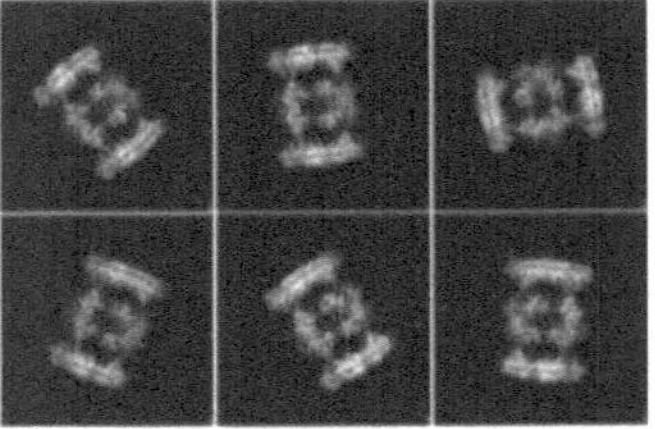

(c) The real valued representations of the reconstructed T20S proteasome signal in Figure 3.22b

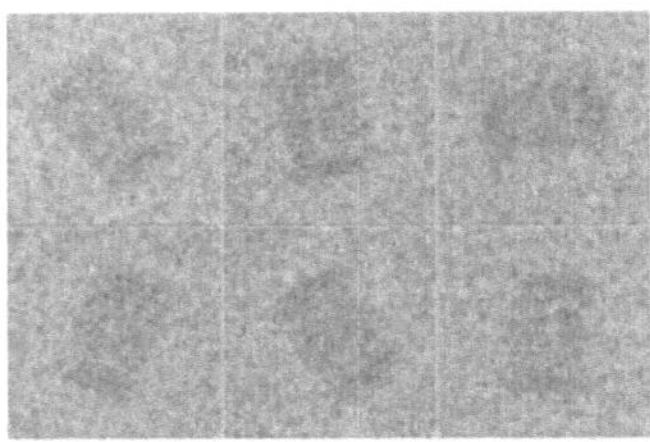

(d) Residuals between the detected in Figure 3.22a and the reconstructed T20S proteasome signal in Figure 3.22b

Figure 3.22: Single particle projections and re-projections of the T20S proteasome All images have the identical optimized parameter set. The cryo-EM data was taken from the experimental data set described in subsubsection 3.3.2.1.

To emphasize again, the recorded single particle projection image is defined as $I_r^{\psi_k}$ (Figure 3.22a) and the re-projection of the reconstructed structure is defined as $I_s^{\psi_k}$ (Figure 3.22b). Here, the index k corresponds to the optimized parameter set ψ of the k-th image in the data.

3.4.1 Mean representation of the reconstructed signal

The software RELION (see subsection 2.4.5) refines the single particle cryo-EM data based on a maximum-likelihood approach (see subsection 2.4.5). Here, the single particle images are assigned to an optimized parameter set of translations and Euler angles based on a probability distribution. The assigned parameters are not distinct. Moreover, the single particle image has a certain likeliness with which it is inserted into the Fourier volume. The 3D protein map is an averaged signal of multiple projection images. Instead of describing the reconstructed signal F_S as in Equation 3.3 the definition of the signal could be altered to

$$F_s = \sum_{i=1}^{T}(F + M_i) \tag{3.15}$$

$$= (T \cdot F + \sum_{i=1}^{T} M_i), \tag{3.16}$$

where T is the number of averaged central sections. Additionally, the reconstructed signal depends on the number of the Fourier shells. With increasing spatial frequency less central sections overlap and hence, less Fourier components are averaged. It results that less signal is averaged and less noise is reduced. Computing the signal-residual ratio, the $QSNR^S$, with respect to the k -th image it is modified to a ratio between the mean reconstruction signal and the k-th recorded image.

$$QSNR_k^S(r, \Delta r) = \frac{\sum\limits_{R \in (r,\Delta r)} \left| \sum\limits_{i=1}^{T}(F + M_i) \right|^2}{\sum\limits_{R \in (r,\Delta r)} \left| (F + M_k) - \sum\limits_{i=1}^{T}(F + M_i) \right|^2} \tag{3.17}$$

$$QSNR_k^S(r, \Delta r) = \frac{\sum\limits_{R \in (r,\Delta r)} |T \cdot F + \epsilon|^2}{\sum\limits_{R \in (r,\Delta r)} |(1 - T) \cdot F + M_k - \epsilon|^2}, \tag{3.18}$$

where the zero-mean Gaussian noise tends to zero, $\lim\limits_{T \to \infty} \sum\limits_{i=1}^{T} M_i = 0$. After the refinement the re-projections are not noise-free (see Figure 3.22b). The reduced noise in these images is described by an $\epsilon \ll 0$. The signal to residual ratio could encounter the scaling factors T and $(1 - T)$ in regards to the signals. This could influence the quality of the QSSNR as the difference does not resemble the true distance between the two signals (see Figure 3.22d). Additionally, the gray values between the images vary so that the distance is influenced.

3.4.2 Normalization of the single particle images

One issue of the projection images and re-projections is the difference in their normalization (see Figure 3.22a, Figure 3.22b, Figure 3.22c). The different gray values of the two images and the position of the signal made it difficult to ensure that the distance is between the two signals. In general, the cryo-EM projection images are standardized. The mean of the images is shifted around zero and the variance of the single particle image equals one. The aim is to equalize the position of the signal information in order to refine the single particle data as entity (see section 2.4.1). However, the re-projections $I_s^{\psi_k}$ have a different image variance compared to the $I_r^{\psi_k}$. Besides, the mean of the re-projection $I_s^{\psi_k}$ is unequal. Most likely the information of the reconstructed signal is shifted with respect to the detected signal and additive noise which are centered around zero. The raw single particle images $I_r^{\psi_k}$ are normalized. The single particle images are defined by

$$I_r^{\psi_k} = f + m, \tag{3.19}$$

where $f \sim \mathcal{N}(\mu_f, \sigma_f)$ and $m \sim \mathcal{N}(\mu_m, \sigma_m)$. The normalization in cryo-EM is a standardization. That means that the mean of the data is subtracted and futher, divide by the variance of the data. The normalized single particle image is represented by

$$I_r^{\psi_k} = \frac{f + m - E[f + m]}{Var[f + m]} \tag{3.20}$$

In general, the variance of two statistical variables is $Var[a + b] = Var[a] + Var[b] + 2cov(a, b)$. Under the assumption of an independence between the protein complex signal and the noise, the normalized image in Normalization of the single particle images is described by

$$= \frac{f + m - E[f] - E[m]}{Var[f] + Var[m]} \tag{3.21}$$

$$= \frac{f - E[f]}{Var[f] + Var[m]} + \frac{m - E[m]}{Var[f] + Var[m]} \tag{3.22}$$

Assuming further that the noise is being distributed with zero-mean and the variance equal to one, the standardized cryo-EM projection image $I_r^{\psi_k}$ is

$$= \frac{f - E[f]}{Var[f] + 1} + \frac{m}{Var[f] + 1}. \tag{3.23}$$

Consequently, the signal of the normalized single particle image is affected by the noise. If the re-projection is not standardized as in Figure 3.22c, the signal is not centered around zero and the distance between both signals is not measurable. If the re-projection is standardized, in the absence of noise the standardization for the re-projection is described by the following equation.

$$I_s^{\psi_k} = \frac{f - E\left[f\right]}{Var\left[f\right]} \tag{3.24}$$

Both standardization lead to a different representation of the signal. Even though the proportion of the signal information within the images is not altered through standardization, they are altered between these two different representations of the signal. The detected signal is affected by the variance of the noise. The distance in Figure 3.22d is not reliable as it is measured between the normalized cryo-EM images and re-projection. One approach to overcome this issue was to define a scaling factor and a translation (see section A.3). Nevertheless, this minimized the distance between the two signal images and hence, has an effect on the residual between these two signals.

The normalizing procedure in RELION differs from the previously described routine. The cryo-EM data processed with RELION is standardized by the approximated mean and variance of the background noise [78]. The resulting normalized single particle image is described by

$$I_r^{\psi_k} = \frac{f + m - E\left[m\right]}{Var\left[m\right]} \tag{3.25}$$

$$= \frac{f}{Var\left[m\right]} + \frac{m - E\left[m\right]}{Var\left[m\right]} \tag{3.26}$$

In RELION the noise is assumed to be distributed with zero-mean and variance equal to σ^2 [70]. Thus, after the standardization the cryo-EM image $I_r^{\psi_k}$ is described as

$$= \frac{f}{\sigma^2} + \frac{m}{\sigma^2}. \tag{3.27}$$

This leads also to a different representation of the signal as compared to a normalized reconstructed signal in Equation 3.24. Consequently, the experiments done with RELION refined cryo-EM data are also challenging in regard to the standardization as the aim is to compute a distance between these signals.

In general, the standardization gives rise to more performance issues. The protein complex underlies a distribution, where the variance does not behave as the noise. The variance of the protein complex depends on its geometric representation. The T20S proteasome, e.g., has a different variance along the one axis (see Figure A.5). Moreover, the assumption of independence between noise and signal is questionable. Structural noise is related to the structure of the protein complex and hence, to the signal of that protein complex [34]. For the standardization of cryo-EM data the covariance between the two image components is not equal to zero and the standardized cryo-EM has a different outcome (see Equation A.1).

After all, standardization is necessary to process the cryo-EM data and to compute a distance between the not normalized single particle image and the reconstructed signal is also complicated. The gray values of the not normalized single particle image range in the thousands so that determining a distance between the recorded and not normalized image and the re-projection is also not possible.

3.4.3   Incorporating the noise

The objective of SPA is to reduce noise in the recorded projection images and hence, enhance the SNR. Averaging over thousand of single particle images improves the SNR. The refined map is an average over all back-projected images. As a result, the re-projection in Figure 3.22b of the reconstructed density map of the protein complex has high an SNR compared to the detected signal in Figure 3.22a. The power of the noise in the original data would dominate the distance between the recorded signal and is re-projected equivalent and hence, the $QSNR^S$ especially in the higher spatial frequencies is influenced. A variety of factors contribute to the reduction of the noise so that Unser *et al.* [76] suggested an empirical model to take the reduced noise in the re-protections into account. This **book** pursued the idea as it has been discussed that the noise component in the projection images is difficult to theoretically model.

The FSC of reconstruction for the noise picking experiment in subsection 4.1.2 estimated the resolution of the reconstructed map of around 9 Å. The FRC of projections came to a similar conclusion for the resolution. However, the non-particle images contain no visible signal in Figure 3.18a and micrographs were recorded with a great certainty that there does not exist a protein complex signal. The $QSNR^S$ in Algorithm between the two images should result in a quality for the numerator and denominator. The residual between the signal of the re-projection and the detected non-particle equals the negative signal and additional noise. The power of the signal and the power of the residual are closely related such that the $QSNR^S$ equals one from a low spatial frequency. The noise reduction is encountered in $QSNR^N$ in Noise reduction. This ratio presented in Figure 3.24 is very small.

Two-sided limit

The theoretical limit of a ratio, where the denominator approaches zero, is the following two-sided limit.

$$\lim_{n\to 0^+} \frac{s}{n} = \infty \qquad \lim_{n\to 0^-} \frac{s}{n} = -\infty, \qquad (3.28)$$

For derived algorithm the denominator in Incorporating the noise can either be replaced by the Γ in Algorithm or the $QSNR^N$ in Noise reduction. The limits of the derived approach in QSSNR are

$$\lim_{QSNR^N\to 0^+} \frac{QSNR^S}{QSNR^N} = \infty \qquad \lim_{QSNR^N\to 0^-} \frac{QSNR^S}{QSNR^N} = -\infty, \qquad (3.29)$$

The QSSNR, which is the basis to compute the FRC of projections, equals the ratio of $QSNR^S/QSNR^N$.

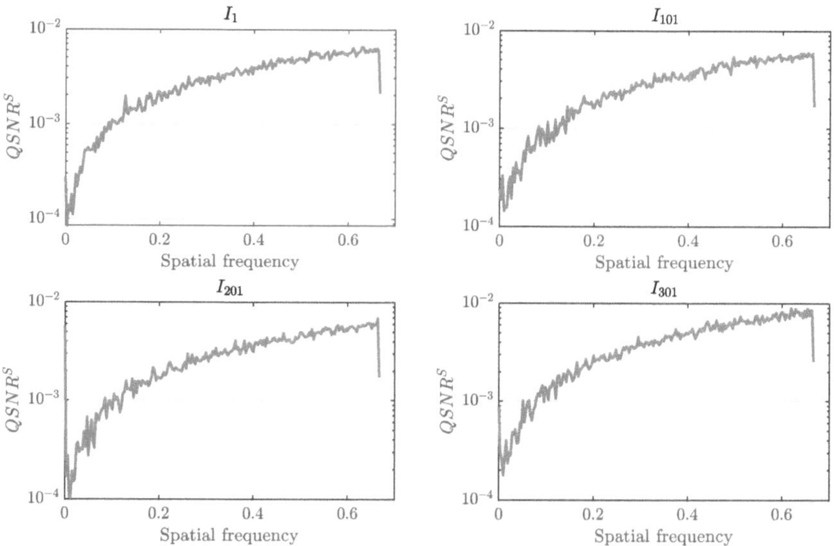

Figure 3.23: $QSNR^S$ **of four different images** Here, the $QSNR^S$ for 4 different distanced signals are plotted in Figure 3.24. These ratios result from the picked non-particles (see subsection 4.1.2). (Logarithmic scales)

The false positive experiment demonstrates the problem between the noise reduction factor $QSNR^N$ in Figure 3.24 and the residual ratio, the $QSNR^S$ in Figure 3.23. The $QSNR^S$ is very small so that the spatial frequency which determines the resolution is equal to the zero frequency and hence, $QSNR^S$ would give an estimated resolution of ∞ Å. However, the $QSNR^N$ is also very small. With the limits given in Incorporating the noise the ratio between $QSNR^S$ and $QSNR^N$ is infinity. This sums up to an average over all measured ratios of $QSNR^S$ to $QSNR^N$. Especially in the lower spatial frequencies, the

QSSNR equals large numbers so that the ratio in Equation 3.14 is greater than one and a higher resolution than the true estimate would be is estimated. Here, the noise reduction factor strengthens the absent signal and further assigns the information to the FRC of projections.

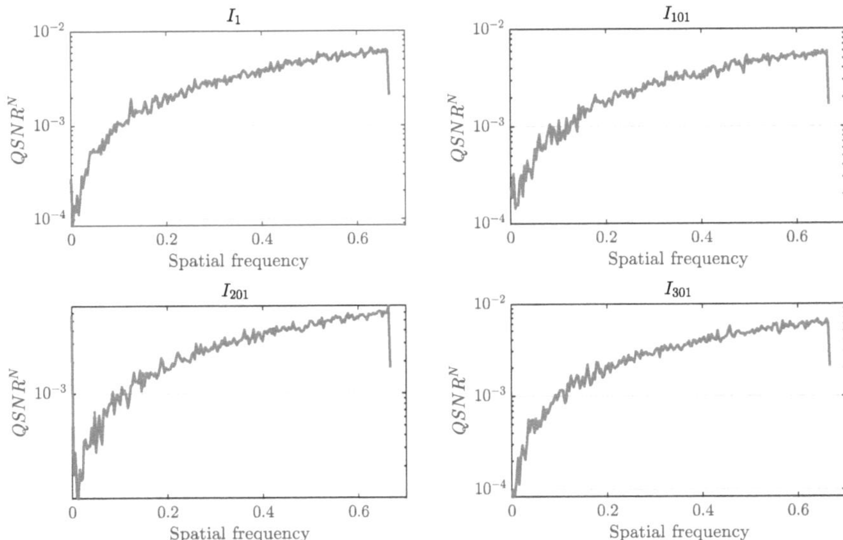

Figure 3.24: $QSNR^N$ **of four different images** Here, the $QSNR^N$ for 4 different projections are plotted. These ratios result from the picked non-particles (see subsection 4.1.2). (Logarithmic scales)

The QSSNR is sensitive to the noise reduction. Especially, in the lower spatial frequencies it boosts the signal to residual ratio so that the FRC of projections is able to estimate a resolution for the experiment. However, dropping the $QSNR^N$ for the noise reduction gives no reliable results for the accurately evaluated resolution data in subsubsection 3.3.2.1. Here, the noise difference between the two signals is significant so that the $QSNR^S$ is dominated by the noise power, which in turn results in a worse estimation of the resolution, which is most likely far off a reasonable estimation.

Chapter 4

Discussion

Structural biology benefits from single particle cryo-EM. With increasing popularity and accessibility more protein structures are reconstructed from cryo-EM data and published. The method is used to study a variety of protein complexes. The imaged protein complexes can be of different sizes, masses and symmetries. Image processing tools have been developed to determine protein complex structure of high resolutions. The single particle projection images are acquired with the TEM and often refined with the state-of-the-art software RELION. The established resolution criterion for cryo-EM data is the FSC. Published cryo-EM structures claim the feature resolution based on this correlation. However, the published cryo-EM density map does not always resemble the true structural features of the protein complex. The noise in cryo-EM data is one of the main drawbacks in reaching atomic resolution and its qualitative data evaluation. This is a consequence of the statistical properties of the noise and the lack of a good SNR of the single particle images, which affect the computational algorithms to refine the data. The three experiments presented in chapter 3 demonstrated the influence of the noise in data processing. The FSC curves shown in Figure 3.3, Figure 3.12 and Figure 3.8 fail to detect the true resolution of the reconstructed protein complexes. The problems related to the overestimation of the resolution by the FSC are discussed here. Further, the **book** aimed to define a validation approach to find a more reliable resolution criteria for the reconstructed maps that includes the noise. The introduced FRC of projections was tested for theoretical and experimental data (see section 3.3). The validation algorithm also led to false assumptions about the protein complex resolution. Indeed, the QSSNR fails to overcome the domination of the noise. This influences the quality of the FRC of projections. Here, the possible issues are discussed.

4.1 From nothing to high-resolution

The cryo-EM image processing tools are sensitive to noise and its behavior. From nothing to high-resolution meant to misuse processing tools to refine noisy cryo-EM projection images to high resolved structures without the corresponding protein signal. One obstacle is the low SNR in the single particle projection images. To increase the SNR multiple similar single particle images are averaged. Therefore, similarly oriented particles need to be identified. However, the high power of noise compared to the signal power makes it difficult to detect similarly oriented projection images. The noise is easily misinterpreted as recorded protein signal because image processing tools cannot precisely distinguish between the recorded signal of the protein complex and the noise. The understanding of the noise effects prevents the misinterpretation and the often resulting overestimation of the resolution. Some effects, e.g. the inaccurate CTF correction (see 3.1.1) or the noise fitting (see 3.1.2), were illustrated in the experiments. The underlying mathematical problem and its interpretation are discussed.

4.1.1 Systematic error within the CTF correction

Two RELION refinements with the identical cryo-EM data were carried out (see section 3.1.1). The difference between these two computations was the on-the-fly CTF correction. To emphasize again, the CTF correction (see subsection 1.3.1) is the processing step, where the data is corrected for some aberrations of the TEM. Hereby, the true image phases which were shifted by defocusing the objective lens due to the weak-phase approximation, are recovered. In one of the two refinements, the defocus parameter δf_{ast} of the CTF was displaced for each single particle projection image. This offset introduced phase errors. The two refined T20S proteasome maps visually differed (see 3.4). Still, the estimated resolution for both maps in Figure 3.3 was high. The question arose if the CTF miscorrection affected the refinement and the estimated feature resolutions. In Figure 4.1 two CTFs with respect to different defocus values are plotted. The variable θ in Figure 4.1 defines the phase shift between the two functions.

Reminder (see 1.3.1)

$$CTF(s) = \sin\left[2\pi\left(\frac{\lambda^3 s^4 C_s}{4} - \frac{\lambda s^2 \delta f_{ast}}{2}\right)\right]$$

The on-the-fly CTF correction is applied on the Fourier transformed single particle image. The parameters of the CTF are specific for each projection image as the CTF has been locally fitted to the power spectrum of the single particle. Under the assumption that the correct CTF values for the Fourier transformed single particle were found, the

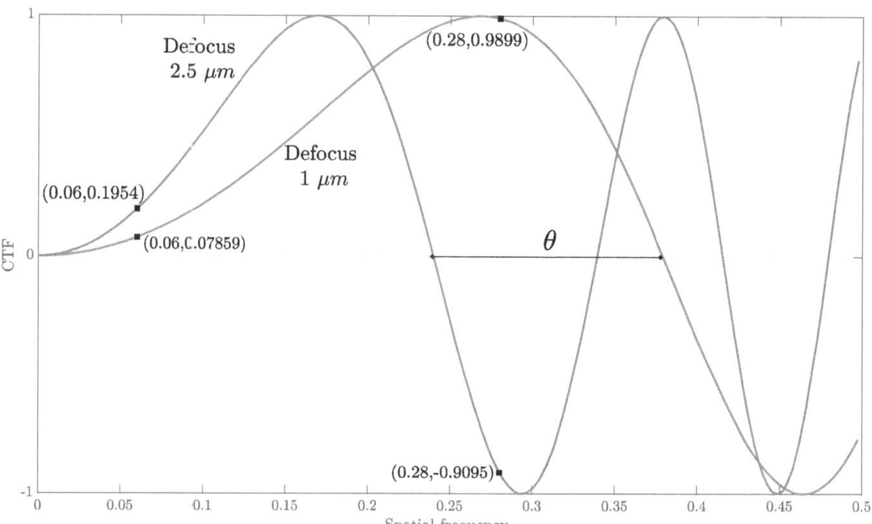

Figure 4.1: Two different CTF corrections Here, two exemplary CTFs are graphed. Both functions are based on the same set of microscopic parameters apart from the defocus. θ marks the difference between the first zero-crossing of the two functions. For comprehension it is assumed that there is no astigmatism present and the defocus of the CTF was optimized. The CTF was defined as the differentiation of the phase from direct to diffracted beam.

phases are optimally recovered. In Figure 3.4b, the map has a high resolution with visibly accurate features of the protein complex. As explained in subsection 3.1.1, the defocus parameters were shifted. It results in visibly different CTFs (see Figure 4.1). The question arose whether the phase error θ propagates into an erroneous reconstruction of the cryo-EM data or attenuates. By applying the 2.5 μm-CTF the phases in the Fourier image do not correspond to the correct scattering information. Consequently, inaccurate phase information is inserted into the 3D Fourier volume. The phase error propagates through the reconstruction since the parameters of the CTF are only calculated during cryo-EM preprocessing. The projection images are repeatedly miscorrected and the refinement of the data leads to an incorrect density map.

The structural representation of the protein complex is false. However, the FSC estimated a high feature resolution for the map. It fails to detect and determine the phase error of the map, even though it is a measurement of phase and amplitude. The failure is subjected to various reasons. One reason is the consistent modification of the CTF parameters in the metadata file. The shift of the defocus was not a completely random process. The displacement had a pattern with respect to the micrographs. It was also initiated before splitting the data such that the phase error is present in both subsets of the data. In the end both half maps of the data refine to a similar geometrical structure. The maps are well correlated as they encounter similar phase errors. Further on, the symmetry of the T20S proteasome contributes to the high correlation of the data measured by the FSC. Here, the T20S proteasome has a symmetry of D7 which means that there a 14 identical subunits. During processing this 14-fold symmetry was imposed. Due to the identical subunits the number of independent voxels within a shell decreases. As a result, both half-maps contain only 1/14 of independent information [46]. The computational algorithms often do not correct for the symmetry.

The effect of miscorrected phases gets worse with increasing spatial frequency. For the low spatial frequencies (see $(0.06, 0.1954)$ and $(0.06, 0.07849)$ in Figure 4.1) the CTF values do not significantly deviate. In the Fourier image the low spatial frequencies correspond to coarse features. Therefore, the overall structure is not significantly affected by the phases displacement. However, with increasing spatial frequencies (see $(0.28, 0.9899)$ and $(0.28, -0.9095)$) the CTF values differ extensively. Detailed features of the protein complex structure are affected as higher spatial frequencies correspond to fast varying information which undergo a more severe phase shift. Since the FSC was not able to detect obvious phase errors, it most likely does not identify smaller differences in the phase information. Particularly, these errors could occur in cryo-EM data processing. The CTF is fitted to the power spectrum of the noisy projection image so that the parameters are only approximated. Smaller deviation between the recorded CTF data and the fitted CTF data cannot be identified and furthermore, not measured by the FSC. Ideally, the validation approach, derived here, would relate the recorded protein complex signal to the reconstructed signal. These two signals deviate. The result would be that the power of the residual signal and the noise is higher than the power of the reconstructed signal. Theoretically, the QSNR in Equation 3.8 should decrease in the lower spatial frequencies. If this occurs, the FRC of projections would estimate a more reliable resolution. However, the algorithm is dominated by the noise. As discussed in section 3.4, defining a distance between the noisy recorded image and the noise reduced image is complicated.

The protein complex is a WPO, so that the recorded phase shifts are small. Further, atomic resolution depends on the high spatial frequency information, which is dominated by the noise. As a consequence, validation of high spatial frequency information is complicated. Due to the noise these fast varying information are not retractable to the recorded signal.

The error introduced during the CTF correction is a systematical error. Systematical means that it is predictable and observable as it is in a relation to the ideal signal. Indeed, it is questionable whether the experiment is a reasonable set up or unlikely in the real world as all defocus values were shifted. Even though a shift including almost all micrographs is not likely, the FSC still was not capable to detect this obvious visible defect within the

3D reconstruction. Running into a phase error related to this setup is unlikely. However, imagining that the phase shift occurs within a small subset of the cryo-EM data, structural details can be reconstructed without being based on the true recorded signal. The phase shift is most likely introduced during preprocessing so that the error propagates into both subsets of the gold-standard refinement. The FSC only measures the consistency between these two subsets, so that this error cannot be determined. Despite the cautious and supposedly independent image processing the error is present in both half maps and hence, consistent.

4.1.2 Overfitting noise

The reference model of a protein complex biases cryo-EM data processing. Using a reference to identify particle on pure noise micrographs resulted in around $80,000$ picked non-particle projection images in the second experiment (see subsection 3.1.2). Furthermore, aligning and matching all these picked images to the reference projections and reconstructing them produced a structure of the protein complex. In Figure 3.7 the reconstructed structure map of the protein complex shows a visible similarity to the reference map. The aim of this experiment was to show that overfitting is present throughout all SPA processing steps. The reference structure, the protein-RNA complex, is reconstructed from noise, where the variation fitted the model variation. Especially the cross-correlation, which measures the similarity between two images in signal processing, is sensitive to overfitting.

Parallel picking-algorithms aim to identify the protein complex signal on the micrograph. Regions are often picked by measuring the CC between the recorded cryo-EM data and templates of reference structure. Thereby, the algorithms fail to distinguish between the protein complex signal and pure noise. A detailed noise-free projection of a reference has a high variance, resulting in template images with a high variance. The recorded signal and the additive noise correlate well to the high variation in the reference projection. As a consequence, pure noise-related regions of the micrograph are assumed to be similar to the reference and hence, are selected. This introduces a model bias. The alignment is also based on the CC (see section 2.4.2). The cross-correlation used to align the recorded cryo-EM data to template images only detects similarities in the values of these images. The projection matching assigns the missing two Euler angles of a specific template image to the picked single particle image. This step is often combined in the reference-based alignment. It does not distinguish between the signal and the noise.

All three algorithms, template-picking, alignment and projection matching, can contribute significantly to overfitting noise. The low SNR of the recorded cryo-EM data can lead to picking, aligning and matching noise, especially if a reference model is used. Shatsky *et al.* [75] states that a 100% chance to correctly align particle images is only given when the data has an SNR of 0.5. With decreasing SNR the probability to correctly match the cryo-EM projection image to the reference decreases. The SNR of recorded cryo-EM data often ranges between 0.1 and 0.3. The SNR of the micrographs cannot be enhanced so that template picking algorithms are subjected to a certain uncertainty. The alignment and projection matching also encounter some uncertainty. To overcome these overfitting problems another approach has been introduced. The maximum-likelihood algorithm as mentioned in subsection 2.4.5 assigns probabilities to the picked particles based on a low resolution reference. In maximum-likelihood algorithm the reference map is filtered to be a smooth structural representation of the protein complex. However, these are also prone to overfitting.

The template picking is done with the identical reference template for all recorded micrographs. Consequently, the model bias is introduced to all cryo-EM projection images. After picking, the data is divided into two subsets for the gold-standard refinement. Even though it is assumed to be an independent computation, the influence of the model propagates into both refinement routines. The FSC shows a decreasing characteristic of the correlation between two reconstructed volumes, so that the resolution can be estimated. Here, the FSC cannot differentiate whether the maps correlated well because of an accurate signal-based structure or because of pure noise. Applying the gold-standard does not protect the data from the template picking model bias. Even more, the FSC depends on the processing style despite the gold-standard refinement. After cryo-EM data processing, the noise is often not statistically independent and does not have to be uncorrelated.

Once again, the FSC is not capable to detect the quality issues of the cryo-EM data. It is not able to identify the discrepancy between the recorded cryo-EM data and the resulting 3D maps because the FSC only considers the two 3D half maps. To stress again, the FSC is no validation tool. The idea of the introduced validation approach was to link the reconstructed data to the recorded data. In theory, the quality issues, seen here, should have been able to be detected. The power spectrum of the reconstructed signal in Equation 3.6 is almost equivalent to the power spectrum of the residual in Equation 3.7. The derived QSNR between these two power spectra should be close to one of the first Fourier shells. Consequently, the QSSNR equals one, the FRC of projections should drop below 0.5. However, the validation fails. Here, the estimated resolution based on the validation approach results from the noise reduction factor (see section 3.2.2). As mentioned in subsection 3.4.3 the noise reduction cannot be neglected to measure the resolution of cryo-EM, e.g. T20S proteasome.

4.1.3   Faking atomic structure

The third experiment demonstrates the model-bias effect on the classification. The variation of the noise for 20% of the projection images within the cryo-EM data fits the variation in the projected faked reference. The detailed structural information of the skull suits the variation in the noise in specific regions of the projection images, whereby the original ribosome structure is still present. Here too, the classification fails to differentiate between the detected signal and the noise. As a result, the algorithm identifies noise. The experiment has an artificial set-up. The density map of a skull does not resemble a true component of a protein complex. However, proteins can bind to another protein complex in order to initiate a biochemical process. These binding processes are often essential for functionality of the protein complex and hence, of interest for *structural biology*.

Clustering algorithms based on probability distribution are known to suffer from over-fitting. The classification implemented in RELION is based on the maximum-likelihood approach. The clustering algorithm is setting the wrong key elements on the data such that it fits noise into the structure. Overfitting scales with exhaustively large numbers of classes to assign the images to. The system is represented by more variables to be optimized than data representing the model. The classification of the 2D projection images in RELION is biased towards the input reference. The likelihood gives the best parameter fit based on the recorded data. However, the maximized estimated parameter set only represents the maximal values present in the data [79]. If the number of classes is too high, the algorithms could define noise as detail to divide single particle into subsets. A consequence could be an over-representation of specific regions. Scheres [80] himself discussed the problematic topic related to the maximum- likelihood algorithm. In general, the maximum -likelihood is also a biased estimator [79].

The FSC in Figure 3.12 is not capable to detect the false reconstruction of the protein complex density. The challenge is again the variation of the noise component. In Figure 3.11 two different enhanced views of the reconstructed density were given. A second issue is the masking effect of the FSC. Using a mask during the refinement compromises the true independence between the two half-set reconstructions [48]. Additionally, the masking of the two maps during post-processing has a filtering effect on the Fourier volumes. Hence, the Fourier components within the shells are affected. The FSC suggests a higher correlation of the Fourier objects with respect to the spatial frequencies. Following the classification the projection images of the reference map and the recorded projection images are cross-correlated such that similar influences exist as present during the refinement. The refinement of 20% of the data that was based on the correlation between reference model projections and the raw data tends to show a positive connection.

The validation of fine features of the protein complexes is complicated. The noise has a strong influence on the distance. It superimposes the details of the skull. The classification was done without an alignment. The validation algorithm would also not be able to identify the quality issues. Even though the projection images of the reconstructed and the recorded signal should differ, the distance was influenced by the power of noise of the data. Furthermore, the classification of the data was done without an alignment. The representation of the noise and the variation of the reconstructed skull could also coincide so that the distance might not differ. The validation could also assume the correctness of the signal. The quality difference between the cryo-EM data and reconstruction, here, is most likely not detectable by the defining a distance.

One of the objectives of SPA is to reach atomic resolution. The dynamical behavior in some parts of the protein complex lead to a non-uniform resolution over the structure map. Regions with higher dynamics often result in less resolution because enhancing the SNR within these regions is complicated. This results from the fact that fewer images are identified to be similar and hence, fewer projection images are averaged. However, the intention is to also reach high resolutions in these regions. References are used to align small features especially in these regions. However, the experiment above showed that whole projection images with different orientations were found in pure noise. Consequently, smaller features, e.g. subunits or side-chains, are more likely to be found in the cryo-EM noise. To detect this overfitting cannot be done with the FSC. The power of the noise dominates the signal power of small features such as side chains. In general, the low SNR makes it often difficult to differentiate between the signal and the noise. Overfitting is more severe within the higher spatial frequencies where the SNR is still low.

4.1.4 Prevent publishing overestimated resolution data

The three experiments in section 3.1 strengthen the proposition that the FSC fails to measure the true feature resolution of the refined cryo-EM protein complex maps. In general, all image processing tools rely on the quality of the recorded data and the user experience. A clean working style of the user is required to process cryo-EM data. The microscope settings during image acquisitions and further while processing need to be precise. The usability advancements in image processing tools allow experienced as well as inexperienced users to process cryo-EM data. The resolution of refined structures should not be claimed by a single number computed with the FSC. Bernard Heymann [81] said „the resolution of the reconstruction is a fair reflection of the errors in alignment". It is strongly recommended to challenge the FSC.

The deposition of the cryo-EM density maps should include the recorded cryo-EM data and detailed description of the image processing routines. Often times, the claimed resolution is estimated by the FSC between masked maps. However, masking routines could have an influence the FSC (see subsubsection 1.4.2.1) [48]. Additionally, to the FSC between the masked 3D maps the FSC between the unfiltered and unmasked 3D structures, which are also computed by RELION, should be computed and deposited.

The resolution of the cryo-EM map is often not a globally defined number. The protein complex is a dynamic object. Local highly dynamic regions of the structure are less resolved than, e.g., a rigid body of the protein complex. However, the resolution of published cryo-EM maps is based on a single estimated number. A measurement to compute local resolutions as it would represent the real world should be considered. The local FSC was introduced [82]. However, it is often not computed or further deposited. This could also prevent false interpretation of detailed information such as shown in the skull experiment subsection 4.1.3.

The outcome of image processing should be visually assessed. The obvious phase error in subsection 3.1.1 can be avoided. Fitting noise into fine structural details, e.g. the skull in subsection 3.1.3, could be prevented by using reference maps with smooth surfaces. High resolution information should not be classified by a reference. However, any reference most likely introduces a model bias. In general, picking and aligning algorithms, which do not use templates, should be preferred. First theoretical attempts to identify the protein complex signal instead of picking single particles are done by auto-correlating the micrographs [83].

In general, validating the resolution of protein complex structures based on the FSC is not trustworthy. Furthermore, the suggesting above are only attempts to prevent the misinterpretation of the cryo-EM data or the overestimation of the resolution. These do not describe a validation of the experimental data. With the advancement of cryo-EM and more protein complex structures going towards atomic resolution, validation methods need to be defined. These should, ideally, verify the resolution based on the recorded signal.

4.2 Validation of noisy cryo-EM data

The FSC does not detect phase errors or aligned non-particle regions. The experiment with the faked reference classified data demonstrates that the FSC cannot identify quality issues of detailed protein complex structures. The FSC is a valid tool to define if the two density maps agree in structural features. However, it does imply if the reconstructed and the recorded signal are qualitatively related. Hence, the aim of this book was to define a validation procedure based on the detected data and the reconstructed map for the qualitative feature resolution assessment. The reconstructed signal is described by the re-projection of the reconstructed protein complex map. A ratio of the re-projection to the distance between the recorded single particle image and the particular re-projection was defined. This ratio describes the power of reconstructed signal to the power of noise and unexplained signal. The idea behind it was that the unexplained signal could be either falsely detected signal or undiscovered signal. As a consequence, the denominator of the ratio should describe a residual of the reconstruction. Ideally, the phase errors in the T20S proteasome map would result in re-projection which differ from the recorded signal. Therefore, especially falsely reconstructed signal would be removed from the resolution assessment. However, the approach cannot be used to determine the true resolution. Difficulties while validating cryo-EM data arise from the interpretation of the noise reduction, the distance between the normalized single particle images and the re-projections as well as the statistical behavior of the noise.

4.2.1 Theoretical vs. experimental noise

The noise as introduced in subsection 1.3.2 is a random process. Cryo-EM noise has been theoretically modeled to be a combination of structural noise, shot noise and detector noise (see section 1.3.3). The established noise model for cryo-EM data is adequate to resolve protein structures to high resolutions. However, in Figure 3.20, the artificial noise images which were generated with MATLAB visibly differ from the recorded cryo-EM data. If validation of the cryo-EM data tries to quantify a residual signal, it also determines the noise in the recorded data. The distance between the noisy single particle images and re-projections also measures the amount of noise. Consequently, it is difficult to determine whether this distance verified the reconstructed signal or is subjected to the power of the noise. Capturing the experimental noise in theoretical noise images is complicated. It is questionable whether the noise model, which is assumed to be independent as well as zero-mean distributed noise, is sufficient. A theoretically correct and in-depth noise model is tough to grasp.

Because of the low SNR in the single particle images, computational algorithm fail to evaluate cryo-EM. Validation, which attempts to examine the true recorded images, is also subjected to these problems. It needs to account for the noise in the recorded data as well as the noise reduction in the refined data. Synthetic data is used to verify computational algorithms which should validate the data. However, artificial noise (section 3.3) cannot reproduce the true behavior of experimental noise sine it is only an approximation of experimental data. Most computation algorithms depend on specific statistical assumptions used to theoretically derive the algorithm. Thus, these algorithms often fail due to the discrepancy between the theoretical noise model and the experimental noise data. The validation is strongly influenced by the power of experimental noise.

Another problem of the noise behavior is the reduction during the refinement. It is a complex process which depends on the data and the processing routines. The differently behaviors in the experiments in subsubsection 3.3.2.3 and subsubsection 3.3.2.1 emphasize this. During the RELION the computation of noise reduction is most likely influenced by the reference model. However, theoretically quantifying the noise reduction is complicated. One possible solution is the design of artificial noise with specific statistical properties. As discussed above, the artificial noise might not describe the full experimental cryo-EM noise. The behavior of the theoretical noise differs from the experimental noise during reconstruction or refinement. Consequently, a ratio of the noise reduction could positively influence the validation of the reconstructed signal. Besides this, the computational costs to experimentally determine the reduction are extensive. Refining theoretical noise images cost computing time and storage space.

In general, data validation is supposed to determine the correctness of the data. Cryo-EM data processing most likely modifies the recorded signal during preprocessing. The cut-out regions of the micrographs are standardized (see section 2.4.1). Even though standardization does not displace information within the single particle image the distribution of the noise impacts the standardization of the recorded single particle image (see subsection 3.4.2). However, not normalizing the recorded cryo-EM data makes it difficult to evaluate thousandth of these single particle projection images as an entity. This complicates to define a distance between the noisy recorded image and the noise reduced re-projection.

4.2.2 Correlation between noisy projection images

Another general issue is the derivation of the relationship between the SSNR and the FSC. It was based on statistical and geometrical assumptions which must not hold true for cryo-EM data. Back in 1974 Bershad & Rockmore [74] derived the general connection between the real space equivalent measurements, the SNR and the NCC. In contrast to this the FSC and SSNR are computed in Fourier space. However, Sorzano *et al.* [73] showed that the link between the SNR and the NCC can be transformed to Fourier space. The second presumption Bershad & Rockmore [74] made, was the independence and zero-mean behavior of the noise and the signal processes. To further simplify, the two processes are assumed to be stationary, band-limited Gaussian processes. If the cryo-EM data is acquired with a sufficient sampling frequency as stated by the Nyquist Shannon Sampling Theorem, the cryo-EM projection image is a sufficient band-limited and uniform representation of the process. Nevertheless, the signal of a protein complex is not a stationary process. In general, a stationary process means that the signal does not undergo a shift in mean and variance with respect to the time. The detected signal in cryo-EM has a spatial dependence. The local mean and local variance depend on the spatial region of the protein complex due to their nature. Within the box of the protein complex, local means and local variances are not constant [73] (see Figure A.5). This contradicting assumption is still present in the derived validation approach. The projection images and the re-projection are not produced by a stationary signal. Due to the nature of a protein complex different regions of the protein complex have different mean values as there are different features and hence, different amounts of atoms present.

Furthermore, Sorzano *et al.* [73] used that the observed band-limited uniform sample values are independent and identically distributed random variables. As a consequence, the dot product of the two sample points of the identical signal equals the sum of the squared power of the signal and the squared power of the noise. This is contradictory as orthogonality and correlation do not imply each other [84]. The cross-term between the noise and the signal component does not have to be zero. The experiments shown in

3.1.2 and 3.1.3 demonstrate how the variation of the noise represents the variation of the signal. An issue is the computation of the real SSNR at small SNR values. Furthermore, as Sorzano *et al.* [73] state, the relationship of FSC and SSNR is questionable. The noise is assumed to be independent in the field of cryo-EM but in reality the noise is dependent. The correlation between two random noise images without processing was presented [47]. The FSC did show a random correlation especially in the low spatial frequencies. With increasing frequencies the correlation between the two noise sections tended to zero.

The relationship between the FSC and SSNR is questioned and hence, does also not hold true for the derived validation approach derived. Consequently, the relationship for the FRC of projections and the QSSNR underlies the similar assumptions and hence, the identical underlying defects. Finally, validation based on the correlation of noisy cryo-EM data lacks certainty of uncorrelated noise. During image processing the noise most likely affects the determination of the protein complex signal. Establishing the amount of noise influence is complex. The correlation cannot be traced back to the signal with a defined certainty. Additionally, the correlation is an indicator of the possibility that two different variables are alike. It cannot identify the effects that caused the protein complex signal. The correlation and causality are two different concepts. Nevertheless, the QSSNR attempts to link a signal that has been caused by scattering and the reconstructed signal. If the SSNR could overcome the noise related issues, it could present a resolution measurement.

4.2.3   Further cryo-EM data validation approaches

Averaging over similar single particle images is one possible approach to enhance the SNR. However, this depends on the identification of similar particles. As seen in this **book**, it is sensitive to the noise. To enhance the SNR of the recorded cryo-EM data other denoising approaches are interesting to be investigated. If denoising of cryo-EM data could be done with more characteristic noise model, the recorded signal could be more straightforward distinguished from the noise. First attempts of denoising cryo-EM data using the geodesic distance [85] have been published.

The distance between the noisy images and the re-projections was difficult to interpret based on a variety of challenges. The distance in high spatial frequencies is biased by noise. Other metrics to measure the distance between the reconstruction and the recorded projection image could be further investigated. However, the relationship between the signal to residual ratio and the FSC would not be improved.

Triangle inequality   Since filtering removes greater amounts of the noise component (see section 2.4.1) the distance between the detected signal and reconstructed signal could be approximated by the distances of the these two images to a Fourier filtered version. The idea is to find the residual between $I_s^{\psi_k}$ and $I_r^{\psi_k}$. A possible distance could be described by the triangle inequality. The inequality could be defined between the projection image distanced to a filtered version of itself $I_{flt}^{\psi_k}$. The second distance defines difference between the re-projection and the identical filtered image. Filtering the projection image 2.4.1 aims to reduce the noise component on $I_r^{\psi_k}$ and therefore, most likely results in a portray of $I_r^{\psi_k}$. The distance between $I_r^{\psi_k}$ and $I_s^{\psi_k}$ is more precise than the triangle distance. This should not be an issue due to the fact that the distance describes the residual between the detected and the reconstructed data.

$$\Delta I = \left| I_r^{\psi_k} - I_s^{\psi_k} \right| \leq \left| I_r^{\psi_k} - I_{flt}^{\psi_k} \right| + \left| I_{flt}^{\psi_k} - I_s^{\psi_k} \right|, \tag{4.1}$$

where $I_s^{\psi_k}$ and $I_r^{\psi_k}$ are the projection and re-projection image with respect to the specific optimized parameter set. The $I_{flt}^{\psi_k}$ is a filtered detected signal image.

Filtering the cryo-EM projection images decreases the noise component. It also modifies the detected signal. The modification depends on the assigned spatial frequencies. However, it would most likely not overestimate the resolution for the reconstructed 3D map. To measure the quality of high resolution features could most likely not be done with this distance.

Chapter 5

Conclusion and Outlook

In all presented experiments, the low SNR in the projection images lead to a false interpretation of the underlying signal and further, a wrong estimation of the resolution. Fitting the noise to signal is one of the main drawbacks. As the refinement algorithms cannot separate signal and noise the variation of the noise will always align well to the variation of the reference signal. Additionally, the miscorrection of the CTF and its resulting displacement of the phases leads to misinterpretation of the data. The advantage of the cryo-EM is the preserved phases of the protein complex, but when these are misplaced the reconstruction of the protein complex can lead to pure nonsense. The FSC as the state-of-the-art measure to define the resolution of a cryo-EM reconstructed protein complex fails to detect the qualitative miss-resolved structures. This correlation measure is sensitive to noise. Three experiments demonstrated how noise affects the image processing algorithms such that the noise information is detected as the signal of a protein complex. Additionally, they underline the statement that the FSC is not a sufficient resolution measure for cryo-EM maps. The aim of this **book** was to define a validation approach based on an SSNR between the detected and the reconstructed signal. This algorithm was derived and verified. The result-ing FRC of projections also failed to detect the true resolution of the resulting structures. Finally, the validation approach is not an effective instrument to estimate the resolution. It was based on similar assumptions related to the connection of the FSC and SSNR which have been shown to be invalid for cryo-EM data. In the end, cryo-EM is still missing a qualitative resolution evaluation.

Different correction factors for the FSC could be implemented. The symmetric factor as introduced in van Heel & Schatz [46] is not encountered in implementation of RELION. The high resolution noise substitution for the computation of the FSC could be implemented as a standard procedure [48]. Furthermore, the resolution validation remains a critical element of the single particle cryo-EM field. Too many ideas claim to be the solution while some are contradicting. To begin with, the structures, which are published, should be questioned based on the knowledge about protein complexes and the recorded data. As presented in the

book, the visual assessment of the refined protein complex maps are possible in the obvious case. Especially, after processing the data the refinement output should be challenged by the user. Oftentimes the lack of knowledge about image processing tools is challenging. As the field of cryo-EM expands, users specialize more into specific research topics. This leads to more users which simply execute the image processing tools. The software packages advance to auto-refine routines and the user must not understand the underlying theory to refine the data. Training the people and giving them an understanding of the tools helps to assess the quality of the refinement. There exist other validation tools like the tilt pair parameter plot for unknown protein complex structures [48]. Furthermore, other structural methods such as the XRC or NMR spectroscopy (see section 1.1) can be used to cross-validate the experimental results. The recorded and unprocessed single particle stack should be published in the data base EMDB. Further investigations on validation tools to verify the data should be done.

In general, the noise model is difficult to establish. The reduction of the noise in the single particle projection image during the refinement is difficult to theoretically construct. The noise is a combination of random processes, which subjected to specific distributions based on their natural appearance. The noise in cryo-EM data is often assumed to be white Gaussian. However, the noise is still present in the reconstructed maps and even dominates the assessment of the data. This gives different possible conclusions. On the one hand there is the possibility of not modeled noise in the image formation process. On the other hand the assumption of zero-mean does not seem to be sufficient. As averaging over hundreds of single particles the noise is not converging to zero but still dominating the higher-spatial frequencies. Moreover, the statistical randomness is often assumed to be Gaussian distributed due to the central limit theorem. This representation might not be sufficient as it does not take all essential components into account. Shot noise is subjected to the Poisson distribution. A further research on different noise models encountering other statistical distributions or a multiplicative noise approach can be done.

Appendix A

Supplements

A.1 Materials and Methods

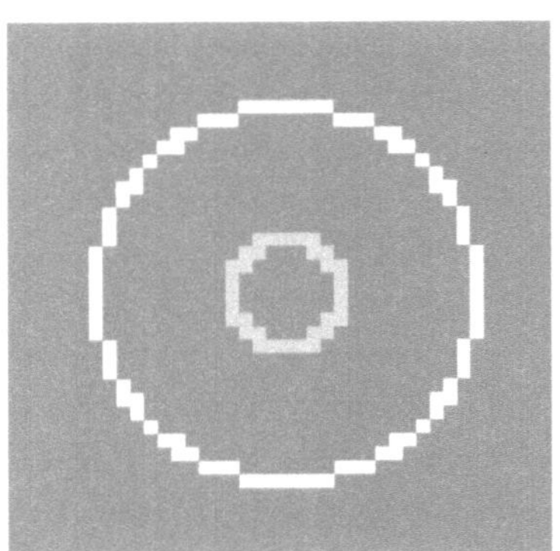

Figure A.1: Fourier rings/shells Here, two Fourier rings with different radii are sketched. In general, with increasing radii the ring has more element. The center point is the DC-component. The with the greatest radii corresponds to the highest spatial frequency. The similar concept holds true for Fourier shells in 3D Fourier space.

A.2 From nothing to high-resolution

T20S proteasome

To evaluate the shift of the defocus values between the two refinements the following MAT-LAB script was written.

```matlab
uiopen('./ctf/wrongDFA.csv',1)
uiopen('./ctf/correctDFA.csv',1)

correctDFA.imgID     = strcat(correctDFA.croppedFromFile,'-',num2str(
    correctDFA.cropCenterX),'-',num2str(correctDFA.cropCenterY));
wrongDFA.imgID       = strcat(wrongDFA.croppedFromFile,'-',num2str(
    wrongDFA.cropCenterX),'-',num2str(wrongDFA.cropCenterY));

wrongDFA             = sortrows(wrongDFA,'imgID','descend');
```

```matlab
correctDFA                  = sortrows(correctDFA,'imgID','descend');
find(correctDFA.imgID ~= wrongDFA.imgID);
find(correctDFA.cropCenterX ~= wrongDFA.cropCenterX);
find(correctDFA.cropCenterY ~= wrongDFA.cropCenterY);
% now both tables show the same single particle at the same table
    position

%%
% unique mircograph id same in both tables
correctDFA.cat  = categorical(correctDFA.croppedFromFile);
correctDFA.M    = findgroups(correctDFA.cat);
wrongDFA.cat    = categorical(wrongDFA.croppedFromFile);
wrongDFA.M      = findgroups(correctDFA.cat);
% here the number of groups equase the vector size- controll var
listMicro       = unique(wrongDFA.croppedFromFile);

find(correctDFA.imgID ~= wrongDFA.imgID);

% unique image id same in both tables
wrongDFA.Num    = (1:989993)';
correctDFA.Num  = (1:989993)';
wrongDFA.imageID_org = correctDFA.imageID;

find(correctDFA.imgID ~= wrongDFA.imgID);

wrongDF     = removevars(wrongDFA, {'imageID','importedFrom','
    cropCenterX','cropCenterY','croppedFromFile','cat'});
correctDF   = removevars(correctDFA, {'imageID','cropCenterX','
    cropCenterY','croppedFromFile','cat'});

%% adding values
wrongDF.diffAngleAll    = correctDFA.rlnAngle - wrongDFA.rlnAngle;
wrongDF.diffdV          = correctDFA.defocusV - wrongDFA.defocusV;
wrongDF.diffdU          = correctDFA.defocusU - wrongDFA.defocusU;

wrongDF.dU_org          = correctDFA.defocusU;
wrongDF.dV_org          = correctDFA.defocusV;
wrongDF.A_org           = correctDFA.rlnAngle;

%% find dU und dV in correct wieder

[LiaTwoA,LocBwoA]           = ismember(wrongDF(:,[1:2]),correctDF
    (:,[2:3]),'rows');

wrongDFloc                  = addvars(wrongDF,LocBwoA);

tmp                         = table2array(correctDF(LocBwoA,:));

wrongDFloc.M_org            = str2double(tmp(:,6));
wrongDFloc.Num_org          = str2double(tmp(:,7));

%% plot abhaengig von sortierung
wrongDFloc = sortrows(wrongDFloc,'M_org','ascend');
figure;
```

```
   plot(wrongDFloc.Num_org,wrongDFloc.Num)
   xlabel('wrongDFloc.Num_org');ylabel('wrongDFloc.Num');
60 figure;
   plot(wrongDFloc.M_org,wrongDFloc.M)
   xlabel('wrongDFloc.M_org');ylabel('wrongDFloc.M');

   %% diff
65 figure;
   plot(wrongDFloc.Num_org,wrongDFloc.diffAngleAll)
   xlabel('wrongDFloc.Num');ylabel('wrongDFloc.diffAngleAll');

   figure;
70 plot(wrongDFloc.Num_org,wrongDFloc.diffdU)
   xlabel('wrongDFloc.Num');ylabel('wrongDFloc.diffdU');

   figure;
   plot(wrongDFloc.Num_org,wrongDFloc.diffdV)
75 xlabel('wrongDFloc.Num');ylabel('wrongDFloc.diffdV');
```

Listing A.1: mapping2.m

The three following graphs illustrate the displacement of the defocus parameter set for each image. The differences in $\delta_{f_u}, \delta_{f_v} \theta_{ast}$ with respect to each picked particle are plotted. All three graphs are row sorted with respect to the defocus difference.

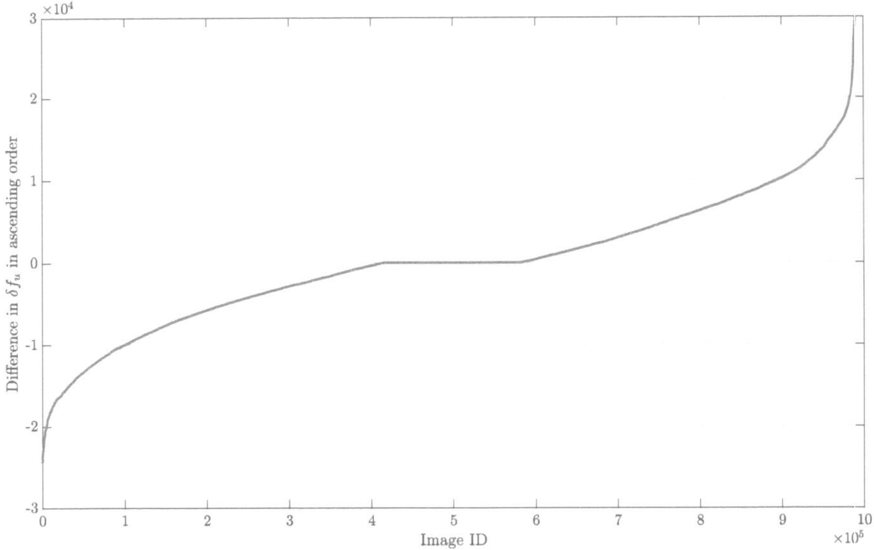

Figure A.2: Defocus difference along the maximum axis of the ellipse

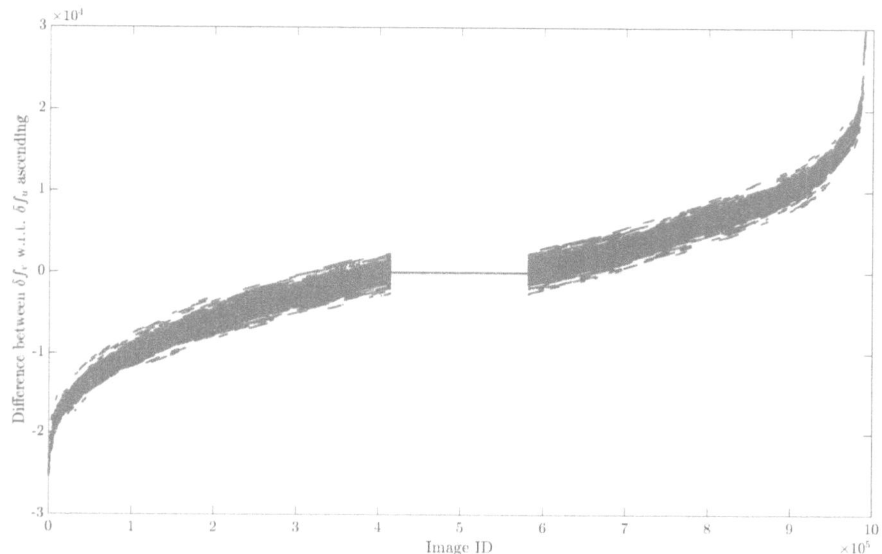

Figure A.3: Defocus difference along the minimum axis of the ellipse

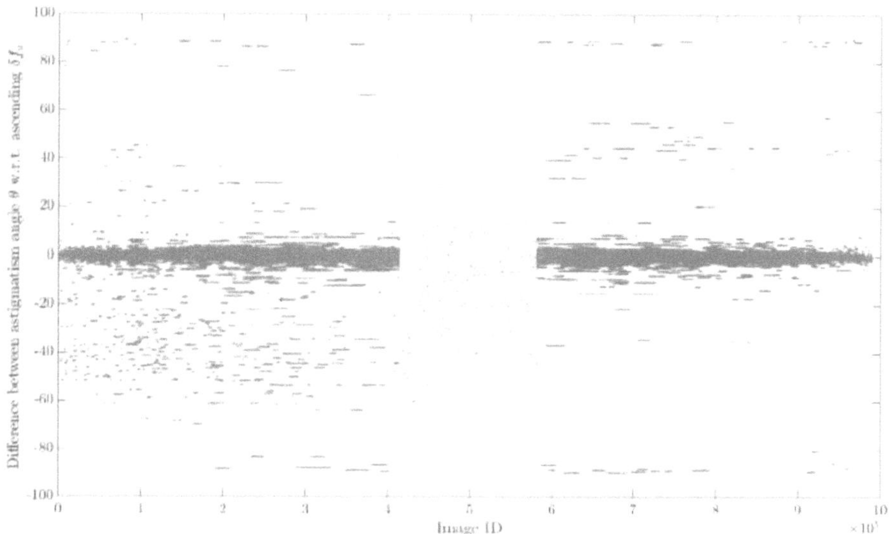

Figure A.4: Difference between the measured angle and the shifted angle

A.3   Results

But the assumption of independence between noise and signal is questionable due to noise related to the structure of the protein complex [34]. For the standardization in cryo-EM is leads to the following equation

$$= \frac{f - E[f]}{Var[f] + Var[m] + 2cov(f, m)} + \frac{m - E[m]}{Var[f] + Var[m] + 2cov(f, m)} \tag{A.1}$$

$$= \frac{f - E[f]}{Var[f] + 1 + 2cov(f, m)} + \frac{m}{Var[f] + 1 + 2cov(f, m)} \tag{A.2}$$

This means that the normalizing produce in SPA is affect by the covariance between the image noise and the signal.

Scaling and translation deviation   As Sorzano *et al.* [68] described the measured signal and the predicted signal is not just disturbed by a random process but also transformed due to physical influences. In general, the Taylor approximation gives a good estimate of the underlying functions. The recorded image I_r^ψ equals an scaled and translated ideal signal I_{id}^ψ.

$$I_r^\psi = a \cdot I_{id}^\psi + b, \tag{A.3}$$

$$\Leftrightarrow$$

$$\frac{1}{a} I_r^\psi - \frac{b}{a} = I_{id}^\psi, \tag{A.4}$$

where I_{id}^ψ is the optimal projection of the recorded image. After the refinement the signal is the sum of various number of recorded images. E.g. the gray values differ between I_r^ψ and the re-projected image I_s^ψ. Under equal assumptions the re-projection and the proejction image are linked by a linear transformation

$$I_r^\psi = a \cdot I_s^\psi + b, \quad \text{where } a := \frac{1}{a}, \ b := \frac{b}{a} \tag{A.5}$$

where $a, b \in \mathbb{R}$. The parameters a, b can be determined by the defining the smallest distance between these two images.

$$\min_{a,b \,\in\, \mathbb{R}} |I_r^\psi - (aI_s^\psi + b)| \tag{A.6}$$

With the least square method the scaling factor a and translation b are determined. By this the signal should be left when solving for least square

$$a = \frac{n^2 \cdot \sum_{i=1}^{I}\sum_{j=1}^{J}\left(I_s^{\psi_k} \cdot I_r^{\psi_k}\right) - \sum_{i=1}^{I}\sum_{j=1}^{J}\left(I_s^{\psi_k}\right) \cdot \sum_{i=1}^{I}\sum_{j=1}^{J}\left(I_r^{\psi_k}\right)}{n^4 \cdot \sum_{i=1}^{I}\sum_{j=1}^{J}\left(I_s^{\psi_k\,2}\right) - \left(\sum_{i=1}^{I}\sum_{j=1}^{J}I_s^{\psi_k}\right)^2}, \tag{A.7}$$

where i, j are the pixel values. The translation b is determined by

$$b = \sum_{i=1}^{I}\sum_{j=1}^{J}I_r^{\psi_k} - a \cdot \sum_{i=1}^{I}\sum_{j=1}^{J}I_s^{\psi_k}, \tag{A.8}$$

where i, j are again the pixel values. With these two parameters a, b the power spectrum of the reconstructed signal becomes

$$\Gamma_k^S(r, \Delta r) = \sum_{R \in (r, \Delta r)} \left| (a \cdot I_s^{\psi_k}(R) + b) \right|^2 \tag{A.9}$$

and the power spectrum with respect to the residual becomes

$$\Gamma_k^N(r, \Delta r) = \sum_{R \in (r, \Delta r)} \left| I_r^{\psi_k}(R) - (a \cdot I_s^{\psi_k}(R) + b) \right|^2. \tag{A.10}$$

Leading to quotient

$$QSNR_i^S = \frac{\Gamma_k^S(r, \Delta r)}{\Gamma_k^N(r, \Delta r)} \tag{A.11}$$

All other equations (3.12), (3.11), (3.13), (3.14) do not change. The computation of the scaling and translation factor is affected by the noise and further affects the QSSNR. There was no reliable computation of these factors possible.

A.4 Discussion

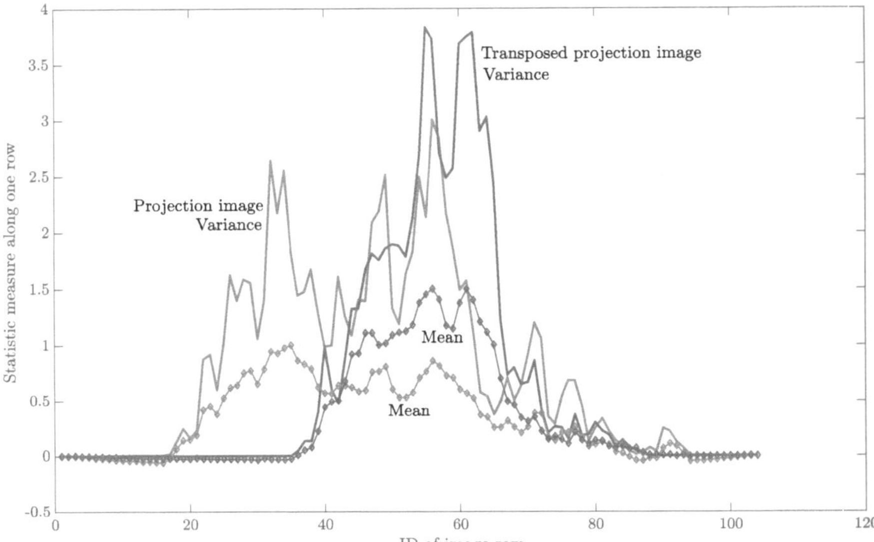

Figure A.5: Local mean and variance of a protein complex Here, the mean and variance of a re-projection of the reconstructed synthetic protein-RNA-complex in Figure 3.14 are computed. The mean was computed along each row, resp. column, of 2D the projection image. The mean as well as the variance differ between the row and column based calculation. The protein is elongated along one axis, which is visible in the local first and second order moments.

Appendix B

MATLAB Coding

Here, the implemented functions and scripts to run the validation algorithm introduced in subsection 3.2.2 are given. All file-IO methods, read and write, were taken from the repository.

Calculate the QSNR of the signal

```matlab
function [SNR, diff] = imageWiseSNR(data, signal, a, b)
% Sabrina Fiedler
% MATLAB 2017b / 2018a

% number of images and dimension
[n,~,N]          = size(data);

for i = 1:N
transformedSignal(:,:,i)   = a(i,1) .* signal(:,:,i) + b(i,1) .* ones(
    n,n);
end

% by linearity of FFT - first subtract and than FFT transfrom
diff              =  double(data - (transformedSignal));
I3                = fftshift(fft2(fftshift(diff)));
I2                = fftshift(fft2(fftshift(double(transformedSignal)
    )));

%% build rings
[x, y]    = ndgrid(-n/2:n/2-1);   % zero at element n/2.
R         = round(sqrt(x.^2 + y.^2));
ring      = zeros(n,n,n/2+1);
nr        = zeros(n/2+1,1);

for i = 1 : n/2+1
ring(:,:,i)    = R == (i-1); % bool value
nr(i,:)        = size(find(ring),1);
end
clear x y R
```

```matlab
    % build a SSNR for each image
30  for i = 1:N
    %% SSNR
    estSignal           = sum(abs(I2(:,:,i)).^2,3);
    rSignal             = ring .* estSignal;
    sumSignal           = squeeze(sum(sum(rSignal,2),1));
35  weightSignal_R      = 1./nr .* sumSignal;

    noise               = abs(I3(:,:,i)).^2;
    estNoise            = sum(noise,3);
    rNoise              = ring .* estNoise;
40  sumNoise            = squeeze(sum(sum(rNoise,2),1));
    weightNoise_R       = 1./nr .* sumNoise;

    SNR(i,:)            = weightSignal_R ./ weightNoise_R;
    end
45  end
    end
```

Listing B.1: QSNR

Calculate the QSNR of the noise, QSSNR and FRC of projections

```matlab
    function [FRC,S,SSNR,SNR] = FRC_from_SNR_SNRdividedByNoise(proj,org,
        noiseOrg,noise,a,b)
    % Sabrina Fiedler
    % MATLAB 2017b / 2018a

5   [n,~,N]             = size(org);

    %% projection and raw images
    [SSNR,~]        = imageWiseSNR(org,proj,a,b);

10  %% FFT of noise
    I3              = fftshift(fft2(fftshift(double(noiseOrg))));
    I2              = fftshift(fft2(fftshift(double(noise))));

    %% define rings
15  [x, y]      = ndgrid(-n/2:n/2-1);   % zero at element n/2 + 1.
    R           = round(sqrt(x.^2 + y.^2));
    ring        = zeros(n,n,n/2+1);
    nr          = zeros(n/2+1,1);

20  for i = 1 : n/2+1
    ring(:,:,i)    = R == (i-1); % bool value
    nr(i,:)        = size(find(ring),1);
    end
    clear x y R

25  % esitmate the SNR of the reconstructed noise to the noise
    for i = 1:N
    estSignal           = abs(I2(:,:,i)).^2;
    rSignal             = ring .* estSignal;
```

```matlab
30  sumSignal          = squeeze(sum(sum(rSignal,2),1));
    weightSignal_R     = 1./nr .* sumSignal;

    estNoise           = abs(I3(:,:,i)).^2;
    rNoise             = ring .* estNoise;
35  sumNoise           = squeeze(sum(sum(rNoise,2),1));
    weightNoise_R      = 1./nr .* sumNoise;

    SNR(i,:)           = weightSignal_R./ weightNoise_R;
    end
40
    % calculate the SSNR of the reconstruction set
    S = max(0,(SSNR./ SNR)-1);
    FRC = S ./ (S + 2);
    end
```

Listing B.2: QSSNR and FSC of projections

Plotting

```matlab
function twoAxis(freqs,A,name1,B,name2,C,name3,filename,titlename)

% Create figure
figure;
5
% Enlarge figure to full screen.
set(gcf, 'Units', 'Normalized', 'OuterPosition', [0, 0, 0.5, 0.7],'
    name',titlename);

ScaleAng = (freqs).^-1;
10
hold on

%% left axis
yyaxis('left')
15  plot(freqs,A,'DisplayName',name1,'LineWidth',1)
    plot(freqs,B,'DisplayName',name2,'LineWidth',1)

% intersection
LineH = get(gca, 'Children');
20  x = get(LineH, 'XData');
    y = get(LineH, 'YData');

Pfrc = InterX([x{2,1};y{2,1}],[freqs(2:end);0.143*ones(size(freqs(2:
    end)))]);
Pfrc05 = InterX([x{2,1};y{2,1}],[freqs(2:end);0.5*ones(size(freqs(2:
    end)))]);
25  Pfsc = InterX([x{1,1};y{1,1}],[freqs(2:end);0.143*ones(size(freqs(2:
    end)))]);
Pfsc05 = InterX([x{1,1};y{1,1}],[freqs(2:end);0.5*ones(size(freqs(2:
    end)))]);

if isempty(Pfsc05)
```

```matlab
     Pfsc05  =   [0,0];
30   end
     if isempty(Pfsc)
     Pfsc    =   [0,0];
     end
     if isempty(Pfrc05)
35   Pfrc05  =   [0,0];
     end
     if isempty(Pfrc)
     Pfrc    =   [0,0];
     end
40
     ylim([-0.2   1.1])
     line([freqs(2),freqs(end)],[0.143,0.143],'Color','k','LineStyle','--'
         ,'DisplayName','0.143')
     line([freqs(2),freqs(end)],[0.5,0.5],'Color','k','LineStyle','-.','
         DisplayName','0.5')

45   ylabel('Linear scale for FSC','Color',[0.47 0.67 0.19]);
     set(axes1,'YColor',[1 0 0]);

     %% second axis

50   yyaxis('right')
     plot(freqs,C,'DisplayName',name3,'LineWidth',1)

     %intersection points
     LineH = get(gca, 'Children');
55   x = get(LineH, 'XData');
     y = get(LineH, 'YData');
     Psnr = InterX([x;y],[freqs(2:end);ones(size(freqs(2:end)))]);

     if isempty(Psnr)
60   Psnr  =   [0,0];
     end

     ylabel('Logarithmec scale for SSNR','Color',[0.47 0.67 0.19]);
     set(gca,'YScale','log','YColor',[0.47 0.67 0.19])
65   line([freqs(2),freqs(end)],[1,1],'Color','k','LineStyle',':','
         DisplayName','1')

     hold off

     % Create title
70   title('FRC (from SSNR) vs FSC','FontSize',24);
     legend

     % Set the remaining axes properties
     set(gca,'FontSize',14,'XTick',freqs(2:10:end),'XTickLabel',round(
         ScaleAng(2:10:end),1));
75   axis(gca,'square');

     % Create xlabel
     xlabel('Angstroem','interpreter','latex');
```

```
80   % textbox with resolution
     dim = [.65 .7 .01 .01];
     str = strcat('Resolution: at 0.5 (at 0.143)','\newline','FSC: ',
         num2str(1/Pfsc05(1,1)),' (',num2str(1/Pfsc(1,1)),')','\newline', '
         FRC: ', num2str(1/Pfrc05(1,1)),' (',num2str(1/Pfrc(1,1)),')','\
         newline', 'SNR at 1: ', num2str(1/Psnr(1,1)));
     annotation('textbox',dim,'String',str,'FitBoxToText','on','FontSize'
         ,16,'LineStyle','none','BackgroundColor',[0.94 0.94 0.94]);

85   savefig(filename);
     end
```

Listing B.3: Plotting

Least square transformation

```
     function [a,b] = transform(proj,raw)
     % Sabrina Fiedler
     % MATLAB 2017b / 2018a

5    proj      = proj(:);
     raw       = raw(:);
     M         = size(proj,1);

     tmp = M * sum(proj .* raw) - sum(proj) .* sum(raw);
10   a   = tmp / (M * sum(proj.^2) - sum(proj).^2);
     b   = 1/M * (sum(raw) - a * sum(proj));
     end
```

Listing B.4: Least square transformation

Main

```
     filename1     = 'einsteinRibo2_a1b0einsteinRibo2_a1b0.mat';
     filename2     = 'einsteinRibo1_a1b0einsteinRibo1_a1b0.mat';

     filename      = strcat('single',filename1);

5    load(filename1, 'freqs');
     load(filename1, 'totalN');

     %% all
10   N = 2 * totalN;

     SNR_total     = cat(1,load(filename1,'SNR'),load(filename2,'SNR'));
     SSNR_total    = cat(1,load(filename1,'SSNR'),load(filename2,'SSNR'));

15   SNR           = cat(1,SNR_total(2).SNR,SNR_total(1).SNR);
     SSNR          = cat(1,SSNR_total(2).SSNR,SSNR_total(1).SSNR);
```

```matlab
FSC             = double(ReadMRC2D('fsc.mrc',1,1));

FSC             = (FSC(:,4));
S_FSC           = 2 * (FSC./(1-FSC));

% per Image
S               = max(0,(SSNR./ SNR)-1);
FRC             = S ./ (S + 1);

FRC             = 1/N * sum(FRC',2);
S               = 1/N * sum(S',2);

twoAxis(freqs(1,2:end),FRC(2:end,1),'FRC',FSC(2:end,1),'FSC',S(2:end
    ,1),'SSNR',strcat(filename,'S.fig'),'S');


% per Image
S1              = max(0,(1/N * sum(SSNR'./ SNR',2))-1);
FRC1            = S1 ./ (S1 + 1);

twoAxis(freqs(1,2:end),FRC1(2:end,1),'FRC',FSC(2:end,1),'FSC',S1(2:end
    ,1),'SSNR',strcat(filename,'S1.fig'),'S1');


% per Image
S2              = max(0,(SSNR./ SNR) -1);
S2              = 1/N * sum(S2',2);
FRC2            = S2 ./ (S2 + 1);

twoAxis(freqs(1,2:end),FRC2(2:end,1),'FRC',FSC(2:end,1),'FSC',S2(2:end
    ,1) ,...
        'SSNR',strcat(filename,'S2.fig'),'S2');
twoAxisSSNR(freqs(1,2:end),FSC(2:end,1),'FSC',S2(2:end,1),'S2'...
        ,0.2.*S_FSC(2:end,1),'S-FSC',strcat(filename,'FSC2SNR.fig'
            ),'Other way around');

%% half
load(filename1,'SNR');
load(filename1,'SSNR');

N               = totalN;
FSC             = double(ReadMRC2D('fsc.mrc',1,1));
S_FSC           = 2 * (FSC./(1-FSC));

% per Image
S               = max(0,(SSNR./ SNR)-1);
FRC             = S ./ (S + 2);

FRC             = 1/N * sum(FRC',2);
S               = 1/N * sum(S',2);

twoAxis(freqs(1,2:end),FRC(2:end,1),'FRC',FSC(2:end,1),'FSC',S(2:end
    ,1),'SSNR',strcat(filename,'S.fig'),'S');
```

```matlab
% per Image
S1             = max(0,(1/N * sum(SSNR'./ SNR',2))-1);
FRC1           = S1 ./ (S1 + 2);

twoAxis(freqs(1,2:end),FRC1(2:end,1),'FRC',FSC(2:end,1),'FSC',S1(2:end
    ,1),'SSNR',strcat(filename,'S1.fig'),'S1');


% per Image
S2             = max(0,(SSNR./ SNR) -1);
S2             = 1/N * sum(S2',2);
FRC2           = S2 ./ (S2 + 2);
```

Listing B.5: main for the ribosome data

```matlab
% Sabrina Fiedler
% MATLAB 2017b / 2018a
%addpath('matlab/')

% Fill in the name
titleName = 't20sPositivExample_90percent_2';

prompt  = 'What is the pixel in A size? ';
pixA    = (input(prompt))

prompt  = 'What number of images in the set? ';
totalN  = (input(prompt))

prompt  = 'What is the dimension of the image? ';
n       = (input(prompt))


% FSC      = (load('FSC_fromStar.mat'));
% FSC      = double(FSC.FSC(:,4));

freqs   = (0:n/2)*1/(n*pixA);

FRC     = zeros(totalN,size(freqs,2));
S       = zeros(totalN,size(freqs,2));
SSNR    = zeros(totalN,size(freqs,2));
SNR     = zeros(totalN,size(freqs,2));

% transform the signal
for i = 1:totalN
% read input data
noise           = double(ReadMRC2D('projection2_noise.mrcs',i,1));
proj            = double(ReadMRC2D('projection2.mrcs',i,1));
noiseOrg        = double(ReadMRC2D('half2_noise.mrcs',i,1));
org             = double(ReadMRC2D('half2.mrcs',i,1));
[a(i,1),b(i,1)] = transform(proj,org);

% validation
[FRC(i,:),S(i,:),SSNR(i,:),SNR(i,:)] = FRC_from_SNR_SNRdividedByNoise(
    proj,org,noiseOrg,noise,a(i,1),b(i,1));
```

```matlab
end


FRC      = 1/totalN  * sum(FRC',2);
summedS  = 1/totalN  * sum(S',2);

filename             = 'SNRdividedByNoise_90percent_2';

csvwrite(strcat(filename, 'SSNR.csv'),SSNR);
csvwrite(strcat(filename, 'SNR.csv'),SNR);
save(strcat(filename,titleName,'.mat'));

% twoAxis(freqs(1,2:end),FRC(2:end,1),'FRC',FSC(2:end,1),'FSC',summedS
    (2:end,1),'SSNR',strcat(filename,'Snr.fig'),titleName);
```

Listing B.6: main

Abbreviations

CC cross-correlation. 32, 54, 95

cryo-EM electron cryo-microscopy. iii, iv, xi, xii, 2, 4–10, 12–14, 17, 19–27, 29, 30, 33, 38, 39, 41, 44, 48, 49, 51–57, 60, 63–66, 70–74, 78, 80–88, 91–96, 98–103, 105, 106, 111

cs-thm central-slice theorem. xi, 27, 38, 49, 50, 56, 67, 70

CTF Contrast Transfer Function. iv, xi, xii, 10–12, 26, 40, 46, 47, 51, 55–59, 78, 92–94, 105

DFT Discrete Fourier Transform. 34

DPR Differential Phase Residual. 52

DQE detective quantum efficiency. 17

EM electron microscopy. 3–5

EMDB Electron Microscopy Data Bank. 6, 7, 106

FFT Fast Fourier Transform. xi, 22, 27, 34–37

FRC of projections Fourier Ring Correlation of projections. iv, xii, xv, 71–73, 76–83, 88–91, 94, 96, 102, 105, 116

FSC Fourier Shell Correlation. iii, iv, xi, xii, 21–26, 30, 52–60, 63, 65–68, 71–73, 76, 77, 79, 80, 91, 93–99, 101, 102, 105

FSC of reconstruction Fourier Shell Correlation of reconstruction. xii, 73, 76–83, 88

FT Fourier Transform. 33, 34

IFFT Inverse Fast Fourier Transform. 34, 50

IFT Inverse Fourier Transform. 34

NCC normalized cross-correlation. 22, 32, 50, 101

NMR spectroscopy Nuclear Magnetic Resonance Spectroscopy. 3–5, 106

PCA Principal Component Analysis. 48

PSF Point Spread Function. 10, 44

QSNR Quality Signal-to-Noise-Ratio. xii, xv, 69–72, 85, 88–90, 94, 96, 115, 116

QSSNR Quality-Spectral Signal-to-Noise-Ratio. iv, xv, 70–73, 76–83, 85, 89–91, 96, 102, 112, 116

RCSB PDB Protein Data Bank. xi, 3–5, 58, 61, 62

RELION REgularized LIkelihood OptiminzatioN. xi, 8, 29, 51, 52, 56, 57, 59, 65, 66, 79, 82, 83, 85, 87, 91, 92, 97, 98, 100, 105

SEM Scanning Electron Microscope. 4

SNR Signal-to-Noise-Ratio. iii, iv, xiii, 7, 10, 14, 15, 17, 20, 21, 24–26, 45–49, 53–55, 61, 73, 80, 84, 88, 91, 92, 95, 96, 98, 100–102, 105

SPA single particle analysis. iii, iv, 4, 6, 8, 13, 17, 20, 29, 45, 48, 52, 56, 88, 95, 98, 111

SSNR Spectral Signal-to-Noise-Ratio. iv, 30, 52–54, 68, 71, 101, 102, 105

STEM Scanning Transmission Electron Microscope. 4

TEM Transmission Electron Microscope. iii, iv, xi, 4, 6–12, 15–17, 19–21, 38–46, 55, 56, 61, 62, 68, 69, 91, 92, 131

WPO weak-phase-object. iv, 9, 26, 94

XRC X-Ray Diffraction Crystallography. 3–5, 106

Nomenclature

N The number of images in the data set. 46

$*$ convolution of two functions $f * g$. 37

$\mathrm{I_r^{\psi_k}}$ the k-th detected single particle projection image. 68, 70, 84, 86, 87, 102, 103

$\mathrm{I_s^{\psi_k}}$ the k-th re-projection image with respect to the reconstruction. 68, 84, 86, 87, 102, 103

$\mathbb{C}$ complex space. 30

$\mathbb{N}$ complex space. 30

$\mathbb{R}$ real space. 30

μm Micormeter is a metrical unit. 11, 93

eV The energy of an electron. 41

keV The accelerating voltage of the TEM. 39, 41

mm Millimeter is a metrical unit. 41

nm Nanometer is a metrical unit. 3, 41

pm Picometer is a metrical unit.. 3, 41

Å Angstroem is a metrical unit with Å equal to $10^{-10}m$. 2, 7, 11, 18, 21, 23–25, 38, 56, 57, 59–61, 63, 65, 66, 74, 76–83, 88, 89

2D Two-dimensional object with $\mathbb{K}^{n \times n}$. xi, 8, 10, 12, 22, 30–32, 34, 35, 37, 38, 41, 46–51, 67, 68, 74, 97, 113

3D three-dimensional map. iii, iv, 6, 8, 10, 12, 13, 15, 17–19, 21–24, 26, 30, 32, 38, 45–52, 56–59, 62, 64–67, 78, 82, 85, 93, 95, 96, 98, 103, 107

MDa Mega Dalton. 6